Ultra-wideband RF System Engineering

This comprehensive summary of the state of the art in ultra-wideband (UWB) system engineering takes you through all aspects of UWB design, from components through the propagation channel to system engineering aspects.

Mathematical tools and basics are covered, allowing for a complete characterization and description of the UWB scenario, in both the time and the frequency domains. UWB MMICs, antennas, antenna arrays, and filters are described, as well as quality measurement parameters and design methods for specific applications. The UWB propagation channel is discussed, including a complete mathematical description together with modeling tools. A system analysis is offered, addressing both radio and radar systems, and techniques for optimization and calibration. Finally, an overview of future applications of UWB technology is presented.

This volume is ideal for scientists as well as for RF system and component engineers working in short range wireless technologies.

Thomas Zwick is a full Professor at the Karlsruhe Institute of Technology (KIT), Germany, and also Director of the Institut für Hochfrequenztechnik und Elektronik (IHE) at the KIT, both roles commencing in October 2007. He has been President of the Institute for Microwaves and Antennas (IMA) since 2008.

Werner Wiesbeck is a full Professor and Distinguished Scientist at the Karlsruhe Institute of Technology (KIT). He has received the IEEE Millennium Award, the IEEE GRS Distinguished Achievement Award, and the IEEE Electromagnetics Award. He is a Fellow of the IEEE, an Honorary Life Member of IEEE GRS-S, a Member of the Heidelberger Academy of Sciences and Humanities, and a Member of the German National Academy of Science and Engineering (acatech).

Jens Timmermann works at Astrium GmbH, Germany, and lectures in electrical engineering at Baden-Wuerttemberg Cooperative State University in Ravensburg, Germany. He is a member of the New York Academy of Sciences.

Grzegorz Adamiuk works on future spaceborne radar systems at Astrium GmbH, Germany. In 2010, he received the prestigious Südwestmetall Award for his scientific work.

Ultra-Wideband RF System Engineering

Edited by

THOMAS ZWICK
Karlsruhe Institute of Technology

WERNER WIESBECK
Karlsruhe Institute of Technology

JENS TIMMERMANN
Astrium GmbH

GRZEGORZ ADAMIUK
Astrium GmbH

CAMBRIDGE
UNIVERSITY PRESS

CAMBRIDGE
UNIVERSITY PRESS

University Printing House, Cambridge CB2 8BS, United Kingdom

Cambridge University Press is a part of the University of Cambridge.

It furthers the University's mission by disseminating knowledge in the pursuit of education, learning and research at the highest international levels of excellence.

www.cambridge.org
Information on this title: www.cambridge.org/9781107015555

© Cambridge University Press 2013

First published 2013

Printed by CPI Group (UK) Ltd, Croydon CR0 4YY

A catalog record for this publication is available from the British Library

Library of Congress Cataloging in Publication data
Ultra-wideband RF system engineering / edited by Thomas Zwick, Werner Wiesbeck, Jens Timmermann, Grzegorz Adamiuk.
    pages   cm
Includes bibliographical references and index.
ISBN 978-1-107-01555-5 (hardback : alkaline paper)
1. Radio – Transmitters and transmission.   2. Ultra-wideband devices.
3. Radio frequency.   4. Ultra-wideband antennas.   5. Ultra-wideband radar.
I. Zwick, Thomas.   II. Wiesbeck, W. (Werner)
TK6553.U48   2013
621.3841′5 – dc23   2013030429

ISBN 978-1-107-01555-5 Hardback

Contents

Contributors

Grzegorz Adamiuk
Astrium GmbH, Germany

Gunter Fischer
Leibniz-Institut für innovative Mikroelektronik, Germany

Xuyang Li
Robert Bosch GmbH, Germany

Christoph Scheytt
Universität Paderborn, Germany

Jens Timmermann
Astrium GmbH, Germany

Werner Wiesbeck
Karlsruhe Institute of Technology, Germany

Thomas Zwick
Karlsruhe Institute of Technology, Germany

Łukasz Żwirełło
Karlsruhe Institute of Technology, Germany

Preface

For many scientists and engineers working in ultra-wideband technology, it seems that the idea of using signals with such a wide instantaneous bandwidth was spread by the US FCC with the accreditation of the frequency band from 3.1 to 10.6 GHz. But, if we look back in history, we find that even the first man-made electromagnetic waves were generated by sparks. Especially famous for electromagnetic research was Heinrich Hertz who, in the 1880s, verified the speed of propagation of electromagnetic waves, their polarization and interaction with objects, and the correct description of these waves by Maxwell's equations at our university in Karlsruhe, Germany. Before this time, electromagnetic waves could only be generated by the aforementioned sparks and were thus ultra-wideband.

Ultra-wideband was banned in the 1920s because it occupied too great a portion of the spectrum and from this point was primarily limited to military applications. This was until 1992 when Leopold Felsen, Lawrence Carin, and Henry Bertoni organized a conference on ultra-wideband, short-pulse electromagnetics in Brooklyn. Our institution, the Institut für Höchstfrequenztechnik und Elektronik (now the Institut für Hochfrequenztechnik und Elektronik) had the privilege of participating in this first conference on ultra-wideband. The topics at the conference were so fascinating that we decided to step into this area. The first research topics were in ground penetration radar, with the idea of detecting anti-personnel mines.

After the first conference a number of other colleagues stepped into the ultra-wideband area and a real ultra-wideband community was established. Since then, in our institution, numerous diploma and master's students, and also PhD candidates, have been working in the ultra-wideband area and its various applications such as radar, communications, localization and medical applications. During this time a detailed knowledge of ultra-wideband electromagnetics, components and system engineering has been developed. As usual, selected topics were published at world-leading conferences and in renowned journals, but most of the detailed results were documented in various internal reports and stored at our laboratory. In 2010 Professor Peter Russer from the Technical University in Munich encouraged us to publish this wide knowledge in a single volume and make it available for the whole community. Our motivation has been to focus on selected topics from the state of the art in ultra-wideband engineering, which will help the reader to understand and develop their ultra-wideband systems and inspire new ideas for further research in this prospective area.

Acknowledgments

The authors and editors would like to thank all those who have supported them with ideas and research projects both before and throughout the writing of this book. In particular, we are grateful to all the diploma, master's and PhD students who contributed a major part to this book through their research and thesis projects. Special thanks go to the German Science Foundation (Deutsche Forschungsgemeinschaft) for their continuous support of our ultra-wideband research work in the UKoLoS program from 2004 to 2012 [167]. Without their funding and encouragement, neither the intensity nor the breadth of our scientific research would have been possible. Within the UKoLoS program, we have enjoyed a particularly fruitful exchange with our colleagues at the Universities of Ilmenau, Berlin, Erlangen-Nürnberg, Duisburg-Essen, Hannover and Ulm. UKoLoS was made a true success by the project's coordinator Professor Dr.-Ing. habil. Rainer Thomä of the Unversity of Ilmenau, whose progressive leadership we have greatly valued.

We would like to acknowledge that the work done by Grzegorz Adamiuk, Xuyang Li, and Jens Timmermann was undertaken during their time at Karlsruhe Institute of Technology. They have since moved to new establishments detailed in the list of contributors.

Finally, we would like to thank our various industrial partners for their support in the development of components and systems for dedicated ultra-wideband applications. We find ourselves in the very fortunate position of being able to rely on close links with German industry, and we do acknowledge that such backing can never be taken for granted. We may not have included all the valuable sponsors, contacts, and sources of ideas from which we have profited in this acknowledgment. Nevertheless, we wish to express our gratitude to all the readers of this book who may feel that they have contributed in one way or another.

Prof. Dr.-Ing. Thomas Zwick
Prof. Dr.-Ing. Dr. h.c. Dr.-Ing. E.h. mult. Werner Wiesbeck
Dr.-Ing. Jens Timmermann
Dr.-Ing. Grzegorz Adamiuk

Notation

Latin symbols

$\tilde{a}$	forward propagating wave amplitude in the frequency domain
a	forward propagating wave amplitude in the time domain
A_{avg}	average level of the peak values of the received pulses
A_{W}	effective antenna area
A_{xt}	average peak level of the noise or cross-talk signal
$af(t, \psi)$	array factor in the time domain
$AF(f, \psi)$	array factor in the frequency domain
b	backward propagating wave amplitude in the time domain
$\tilde{b}$	backward propagating wave amplitude in the frequency domain
B	observation point
B	signal bandwidth
B_{a}	absolute bandwidth
B_{r}	relative bandwidth
BW	impedance bandwidth ($S_{11} < -3$ dB)
C	Shannon channel capacity
C	complex radiation pattern
d	distortion
d	distance between array elements
d	distance
D	antenna dimension
D	directivity
$\mathbf{e}$	electric field strength vector in the time domain
E_{b}	bit energy
$\mathbf{E}$	electric field strength vector in the frequency domain
$\mathbf{E}^{\mathrm{S}}$	scattered electric field strength
f	frequency
f_{c}	geometric center frequency
f_{l}	lower frequency bound
f_{u}	upper frequency bound
f_{PRF}	pulse repetition frequency
F	fidelity
g_0	Green's function of free space

g_T	transient gain
G	antenna gain
G_{Rx}	antenna gain of receiver
G_{Tx}	antenna gain of transmitter
h	impulse response in the time domain
h_{Rx}	height of receiver over ground
h_{Tx}	height of transmitter over ground
H	transfer function in the frequency domain
H_G	generator voltage transfer function
H_{ges}	overall transfer function
H_{Klm}	port voltage transfer function
$\mathbf{h}$	full polarimetric impulse response in the time domain
$\mathbf{H}$	full polarimetric transfer function in the frequency domain
$\mathbf{H}_{oc}$	effective antenna height related to open circuit voltage
i	counter
i	current in the time domain
I	current in the frequency domain
j	imaginary unit $j = \sqrt{-1}$
$\mathbf{j}$	current density in the time domain
$\mathbf{j}^{\delta}$	current density in the time domain related to a Dirac excitation
$\mathbf{J}$	current density in the frequency domain
k	wave number
K	constant of Wiener filter
l	length
$L_{FS}(f)$	free-space attenuation
$L_{FS,UWB}(f)$	free-space attenuation of UWB signal
$L_{two\text{-}path}(f)$	free-space attenuation of two-path model
m	counter
M	number of positions
n	counter
N	number of elements
N	noise power
N_0	noise spectral density
N_m	number of propagation paths
N_{ortho}	number of orthogonal pulses
N_{TH}	number of time-hopping time slots
O	center of origin
O_Q	center of radiation
p	polarimetric matching
p	peak value of impulse response
$p(t)$	pulse shape in the time domain
P_{loss}	loss
P_{rad}	radiated power

P_{refl}	reflected power
P_{Rx}	total receive power
P_{Tx}	total transmit power
Q	quality factor
Q	error function
r	radius, distance to transmitting antenna
$\tilde{r}$	reflection coefficient in the frequency domain
r_{A}	radius of smallest sphere that can contain the antenna
r_{CCF}	cross-correlation function
r_{Q}	distance from center of origin to center of radiation
r_{TxRx}	distance between transmitter and receiver
R	data rate
S	signal power density
S	signal power
S_{11}	input reflection coefficient
S_{21}	transmission coefficient
S_{12}	feedback coefficient
S_{22}	output reflection coefficient
$[\mathbf{S}]$	scattering matrix
S/H	sample and hold
t	time
T	duration in time or duration of a period
T	temperature
T_0	time step
T_{p}	pulse duration
T_{PPM}	PPM time offset
T_{TH}	length of time-hopping time slot
$\mathbf{T}_i$	transmission coefficient of polarimetric propagation path
u	voltage in the time domain
U	voltage in the frequency domain
U_{BP}	bandpass signal
U_{G}	generator open circuit voltage
U_{oc}	receiving antenna open circuit voltage
V	volume
w_i	weighting coefficient used in the time domain
W_i	weighting coefficient used in the frequency domain
Z	impedance
Z_{C}	characteristic impedance
Z_{G}	generator impedance
Z_{L}	load impedance

Greek symbols

α	fraction of peak value used in ringing definition
α	attenuation coefficient
β	phase coefficient
γ	absolute value of reflection coefficient of second path
γ	complex propagation constant
δ	Dirac impulse
Δl	path length difference
ΔR	range resolution
ε	permittivity
ε_0	free-space permittivity
ε_r'	real part of relative permittivity
ε_r''	imaginary part of relative permittivity
η	efficiency
θ	elevation angle in spherical coordinates
Θ_{mb}	main beam direction
λ	wavelength
λ_0	free-space wavelength at center frequency
ξ	polarimetric ratio
ρ	cross-correlation coefficient
σ	conductivity, standard deviation of the noise signal
σ	conductivity of medium
σ_G	standard deviation of G
$\sigma_{\tau_{\mathrm{G}}}$	standard deviation of group delay
τ	time duration or delay
$\overline{\tau}_{\mathrm{D}}$	average delay time
τ_{DS}	delay spread
τ_e	true time delay increment
τ_{FWHM}	duration of full width at half maximum
τ_{g}	group delay
τ_{r}	duration of ringing
τ_{rad}	antenna signal delay from port to far field port
τ_{TOF}	time of flight
ϕ	phase of reflection coefficient of second path
Φ	electric potential in the time domain
φ	phase angle
ψ	azimuth angle in spherical coordinates
ψ_{mb}	main beam direction
ω	angular frequency
Ω	steradian

Operators and mathematical symbols

r	scalar
$\mathbf{r}$	vector
$\mathbf{r}^{\mathrm{T}}$	vector $\mathbf{r}$ transposed
$\hat{\mathbf{r}}$	unit vector parallel $\mathbf{r}$
$\hat{\mathbf{r}}_\theta$	local base unit vector in θ-direction
$\hat{\mathbf{r}}_\psi$	local base unit vector in ψ-direction
$\hat{\mathbf{r}}_r$	local base unit vector in r-direction
	with $\hat{\mathbf{r}}_r = \hat{\mathbf{r}}$ in spherical coordinates
$\hat{\mathbf{r}}_z$	local base unit vector in z-direction ($\theta = 0$)
$|\mathbf{r}|$	absolute value of $\mathbf{r}$
$\mathbf{r}_1 \cdot \mathbf{r}_2$	scalar product of $\mathbf{r}_1$ and $\mathbf{r}_2$
$\mathbf{r}_1 \times \mathbf{r}_2$	vector product of $\mathbf{r}_1$ and $\mathbf{r}_2$
$r_1 * r_2$	convolution integral of r_1 and r_2
$\mathbf{r}_1 * \mathbf{r}_2$	convolution integral analog to a scalar product of $\mathbf{r}_1$ and $\mathbf{r}_2$
$[\mathbf{r}]$	matrix
$[r]$	physical unit of r
$\Re\{\cdot\}$	real part
$\mathbb{R}^3$	3D vector space
$\mathbb{R}^3 \backslash V_\mathrm{A}$	$\mathbb{R}^3$ without the volume V_A
$\mathbf{H}^+$	analytic signal of $\mathbf{H}$
$\mathbf{H}^*$	conjugate complex of $\mathbf{H}$
$\mathbf{H}^{\mathrm{T}}$	transposed matrix of $\mathbf{H}$
$\mathcal{H}\{\cdot\}$	Hilbert transform
$\|\mathbf{H}\|_p$	p-norm of $|\mathbf{H}|$
$\overline{G}$	integral average of G over frequency
$\|h(t)\|_2$	2-norm of $h(t)$
$\angle H$	phase angle of H
det	determinate
div $\mathbf{a}$	divergence (sources) of $\mathbf{a}$
exp	exponential function
grad a	gradient of a
ln	natural logarithm
log	logarithm to the base 10
max	maximum
min	minimum
rot $\mathbf{a}$	rotation (curls) of $\mathbf{a}$
sup	supremum
∞	infinity
$\propto$	proportional

General indices

A, ant	antenna
ar	array
BP	bandpass
co	copolarisation
feed	feed
FF	far field
FS	free space
G	generator
h	horizontal
L	load or line
mb	main beam
Mod	model
PC	propagation channel
r	radial
ref	reference
rel	relative
Rx	receiver
Tst	test
Tx	transmitter
v	vertical
x	cross-polarisation
xt	cross-talk
+	forward propagating wave
−	backward propagating wave

Constants

c_0	speed of light in vacuum: 2.997925×10^8 m/s
C	Euler–Mascheroni constant: $0.577\ldots$
e	Euler number: $2.718\ldots$
ε_0	permittivity of vacuum: 8.854×10^{-12} As/(Vm)
k	Boltzmann constant
μ_0	permeability of vacuum: $4\pi \times 10^{-7}$ Vs/(Am) $\approx 1.257\ldots \times 10^{-6}$ Vs/(Am)
π	ratio of circumference to diameter of a circle $3.1415\ldots$
Z_{F0}	wave impedance in vacuum: $Z_{F0} = \sqrt{\frac{\mu_0}{\varepsilon_0}} \approx 377\,\Omega$

Acronyms

3D	3-dimensional
ACR	auto-correlation receiver
ADC	analog–digital converter
ADS	advanced design system
AF	array factor
AIR	antenna impulse response
AoA	angle of arrival
AoD	angle of departure
AUT	antenna under test
AWGN	additive white Gaussian noise
BAN	body area network
BB	base band
BBH	broadband horn antenna
BER	bit error rate
BJT	bipolar junction transistor
BPSK	binary phase shift keying
bs	boresight
BS	base station
CAD	computer aided design
CDF	cumulative density function
CMOS	complementary metal oxide semiconductor
CPW	coplanar waveguide
CR	correlation receiver
CSL	coupled slotline
CT	computed tomography
CW	continuous wave
DAC	digital–analog converter
DC	direct current
DCO	digitally controlled oscillator
DFG	Deutsche Forschungsgemeinschaft (*German Research Foundation*)
DFT	discrete Fourier transform
DLL	delay-locked loop
DoA	direction of arrival

DoD	direction of departure
DOP	dilution of precision
DS	delay spread
DUT	device under test
ECC	Electronic Communications Committee
ECG	electrocardiogram
EF	element factor
EIRP	equivalent isotropically radiated power
EM	electromagnetic
ESD	electrostatic discharge
EuMA	European Microwave Association
EurAAP	European Association on Antennas and Propagation
FBW	fractional bandwidth
FCC	Federal Communications Commission
FD	frequency domain
FDTD	finite difference time domain
FFT	fast Fourier transform
FIR	finite impulse response
FPGA	field programmable gate array
FR	flashing receiver
FWHM	full width at half maximum
GDOP	geometrical dilution of precision
HDOP	horizontal dilution of precision
HPIB	Hewlett Packard interconnect bus
IC	integrated circuit
ICU	intensive care unit
IEE	Institution of Electrical Engineers, part of IET since 2007
IEEE	Institute of Electrical and Electronics Engineers
IET	Institution of Engineering and Technology
IFFT	inverse fast Fourier transformation
IHE	Institut für Hochfrequenztechnik und Elektronik at KIT
IHP	Innovations for High Performance Microelectronics (Research Institute of the Leibniz Association in Frankfurt/Oder, Germany)
IIR	infinite impulse response
INS	inertial navigation system
IR	impulse response
IR-UWB	impulse radio ultra-wideband
ISI	inter-symbol interference
ISO	International Organization for Standardization
KIT	Karlsruhe Institute of Technology
lhc	left-hand circular
LMS	least mean square
LNA	low-noise amplifier

LO	local oscillator
log-per	logarithmic periodic antenna
LOS	line-of-sight
LPDA	logarithmic periodic dipole array
LR	left–right
LTI	linear time invariant
LU	lower–upper
LUT	look-up table
MAC	multiple access
MBM	measurement data-based model
MIKON	International Conference on Microwaves, Radar & Wireless Communications
MIMO	multiple input multiple output
ML	maximum length
MOSFET	metal oxide semiconductor field-effect transistor
MRI	magnetic resonance imaging
MU	mobile unit
MW	microwave
NESP	normalized effective signal power
NLOS	non-line-of-sight
OFDM	orthogonal frequency division multiplexing
OOK	on–off keying
OPM	orthogonal pulse modulation
PA	power amplifier
PCB	printed circuit board
p-cg	p-center of gravity
PDF	probability density function
PDP	power delay profile
PEG	polyethylene glycol
PG	pulse generator
PGA	programmable gain amplifier
PGC	programmable gain control
PGEN	pulse generator
PLL	phase locked loop
PN	pseudo noise
PPM	pulse position modulation
PRF	pulse repetition frequency
PSD	power spectral density
PVC	polyvinyl chloride
PVT	process, voltage and temperature
RAIM	receiver autonomous integrity monitoring
RCM	range comparison method
RCS	radar cross-section

RF	radio frequency
RFID	radio frequency identification
rhc	right-hand circular
RMS	root mean square
RSS	received signal strength
Rx	receiver
SAR	synthetic aperture radar
SER	symbol error rate
SIB	system interconnect bus
SISO	single-input single-output
SMA	sub-miniature plug type A
SNR	signal-to-noise ratio
SPI	serial peripheral interface
SRD	step recovery diode
STR	signal to threshold ratio
SVR	support vector regression
TD	time domain
TDC	time-to-digital converter
TDMA	time division multiple access
TDoA	time difference of arrival
TEM	transversal electric magnetic
TH	time-hopping
ToA	time of arrival
TOF	time of flight
TR	transmitted reference
TTD	true time delay
TWR	two-way ranging
Tx	transmitter
US	United States
UWB	ultra-wideband
VCO	voltage controlled oscillator
VDOP	vertical dilution of precision
VGA	variable gain amplifier
VNA	vector network analyzer
VSWR	voltage standing wave ratio
WBAN	wireless body area networks
WLAN	wireless local area network

1 Introduction

Jens Timmermann and Thomas Zwick

This book concentrates on UWB RF systems. In the analog RF frontend, a high relative bandwidth and not necessarily a high absolute bandwidth poses new challenges to the RF system design. We are therefore concentrating on the lower frequency range of around 1–10 GHz where a large variety of system concepts are under investigation worldwide. A short list of typical applications envisioned by researchers and companies is:

- high data rate, short range applications: typically portable devices and built into antenna systems in consumer electronics or into access point infrastructure
- low data rate, wider range, eventually combined with a ranging application: small portable devices (also wearable systems etc.) combined with integrated antenna system for the accesspoint infrastructure
- low data rate and a high number of users (sensor networks), typically small integrated antennas
- medical imaging for diagnostics, radar systems in combination with antenna arrays
- localization for industrial, medical and commercial applications
- high resolution radar for various applications (e.g. mine detection, through-wall imaging, material inspection).

This chapter provides the definition of UWB signals and regulatory aspects.

1.1 Definition of UWB signals

An ultra-wideband signal is either a signal with a simultaneous bandwidth B that satisfies the condition

$$B \geq 500 \text{ MHz} \tag{1.1}$$

or it is a signal with a relative (=fractional) bandwidth f_r larger than 20% [46]. The relative bandwidth is defined as

$$B_r = \frac{f_u - f_l}{f_c}. \tag{1.2}$$

Table 1.1 US FCC regulations: limits of the PSD for indoor applications [46].

Frequency range GHz	PSD dBm/MHz
Below 0.96	−41.3
0.96–1.61	−75.3
1.61–1.99	−53.3
1.99–3.1	−51.3
3.1–10.6	−41.3
>10.6	−51.3

In this equation f_u and f_l denote the upper and lower frequencies at which the power spectral density is 10 dB below its maximum. f_c is the center frequency:

$$f_c = \frac{f_u + f_l}{2}. \tag{1.3}$$

1.2   Worldwide regulations

The maximum emission levels of UWB devices are defined by specific UWB regulations. Different countries have released regulations (e.g. the National Frequency Plan) which cover the following points:

- applications of UWB technology (indoor, outdoor, portable, fixed installed)
- allocated frequency ranges
- maximum emission levels: power spectral density (PSD) in terms of equivalent isotropically radiated power (EIRP)
- techniques to mitigate (reduce) possible interference caused by the UWB device.

UWB regulations have been released by the United States, Europe, Japan, Korea, Singapore and China. The US Federal Communications Commission (FCC) was the first authority worldwide that released UWB regulations in February 2002 [46]. According to the FCC regulations, the usable frequency range for UWB indoor applications is between 3.1 and 10.6 GHz. The emission limits are defined in Table 1.1.

In Europe, the regulations have been available since March 2006 [44]. They describe respective levels for indoor applications. The (technically) usable frequency range in the EU is allocated to two bands: 4.2–4.8 and 6–8.5 GHz. However there are some constraints on the first band – a mitigation technique has to be used. Without mitigation, the requirement is −70 dBm/MHz rather than −41.3 dBm/MHz. The maximum emission levels resulting from the European regulations are summarized in Table 1.2.

For completeness, Table 1.3 lists the technically usable frequency ranges for all the countries that have released UWB regulations. The maximum emission level is −41.3 dBm/MHz in all cases. UWB signals present an ultra-large bandwidth, which can, for example, be used to realize very high data rates (> 100 Mbit/s). It is also possible to

Table 1.2 ECC regulation: limits of the PSD for indoor applications [44].

Frequency range GHz	PSD dBm/MHz
Below 1.6	−90.0
1.6–2.7	−85.0
2.7–3.4	−70.0
3.4–3.8	−80.0
3.8–4.2	−70.0
4.2–4.8	−70.0/−41.3
4.8–6.0	−70
6.0–8.5	−41.3
8.5–10.6	−65
>10.6	−85

Table 1.3 Technically usable frequency range for countries with UWB regulation [16, 34, 41].

Nation	1st frequency range GHz	2nd frequency range GHz
USA	3.1–10.6	–
Europa	4.2–4.8	6.0–8.5
Japan	3.4–4.8	7.25–10.25
Korea	3.1–4.8	7.2–10.2
Singapore	4.2–4.8	6.0–9.0
China	4.2–4.8	6.0–9.0

make use of the ultra-fine time resolution (with applications in localization and imaging). However, one has to consider that the total emitted power has to be very low to fulfill the regulatory aspects: the limitation to −41.3 dBm/MHz between 3.1 and 10.6 GHz results in a total transmitted power of only 0.56 mW for the FCC mask. For the European mask, the value is even smaller. As a consequence, commercial UWB transmission is limited to short range applications. To exploit the technically usable UWB frequency range, two different approaches are possible:

- Approach 1: Transmission based on ultra-short pulses, which cover an ultra-wide bandwidth (also called impulse radio).
- Approach 2: Transmission based on Orthogonal Frequency Division Multiplexing (OFDM), where the total UWB bandwidth is subdivided into (and exploited by) a set of broadband OFDM channels.

Considering Approach 1, it is desirable to make use of pulses that show a nearly constant spectrum in the technically usable frequency range in order to maximize the overall signal power with regard to the emission regulation. On the other hand, a cost-efficient solution may be to use classical pulse shapes that are easy to generate, but not very efficient with respect to the exploitation of the mask (hence degraded signal-to-noise ratio and degraded performance). A pulse shape that is easy to generate is the Gaussian monocycle or one of its derivatives. In general, impulse radio transmission does not make use of a carrier, which means that the signal is directly radiated via the UWB antenna. Impulse radio therefore has the potential of realization with reduced complexity in comparison with traditional narrowband transceivers.

For Approach 2, the spectral mask can be exploited more efficiently. On the other hand, OFDM transmission leads to increased complexity in terms of signal processing. The overall power consumption due to the increased signal processing may be higher compared to impulse radio transmission. The selection between the two approaches depends on the application and will be a case by case decision.

2 Fundamentals of UWB radio transmission

Jens Timmermann and Thomas Zwick

UWB is an umbrella term that mainly indicates that a very large absolute bandwidth ($B_a > 500$ MHz) or a very large relative bandwidth ($B_r = 2[f_u - f_l]/[f_u + f_l] > 0.2$) in the RF spectrum is used instantaneously by the system. With this definition, no special purpose or application and no special modulation is defined but it implies that the components of the system must be capable of handling this wide spectrum. As already mentioned in the previous chapter for RF frontends, on the whole it is the relative bandwidth that poses new challenges, so system aspects for a very large relative bandwidth are mainly discussed here. This chapter provides a mathematical description of the UWB radio channel including the antennas and measures to characterize the UWB performance of the analog frontend, including the radio channel in the frequency domain (FD) and in the time domain (TD). The chapter presents two methods to exploit an ultra-wide bandwidth: the transmission of short pulses in the baseband (impulse radio transmission), and the transmission by a multi-carrier technique called orthogonal frequency division multiplexing (OFDM). For impulse radio, the most common pulse shapes are introduced together with methods to generate them. Finally, modulation and coding techniques are considered as well as basic transmitter and receiver architectures. The coordinate system is given in Fig. 2.1.

2.1 Description of the UWB radio channel

Typically, narrow-band systems are described in the frequency domain. The characteristic parameters are then assumed to be constant over the considered bandwidth. Due to the large relative bandwidth to be considered for UWB systems, the frequency-dependent characteristics of the antennas and the frequency-dependent behavior of the channel must be taken into account. On the other hand, UWB systems are often realized in an impulse-based technology, so a time domain description might be advantageous as well [154]. Hence there is a requirement for both a frequency domain representation and a time domain representation of the system description.

2.1.1 Time domain and frequency domain

Ultra-wideband signals can be represented both in the TD and in the FD. In the TD, the signal is described as a function of time. The Fourier transformation of a signal in

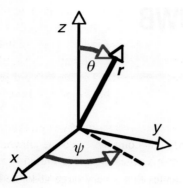

Figure 2.1 Coordinate system for UWB link and antenna characterization.

the TD leads to a representation in the FD, called the spectrum. The inverse Fourier transformation of the spectrum leads to the representation in the TD. Mathematically speaking, this means: to perform a (continuous) Fourier transformation, the continuous complex signal $f(t)$ ($\Re \mapsto C; t \mapsto f(t)$) in the TD has to fulfill the following condition:

$$\int |f(t)|dt < \infty. \tag{2.1}$$

This condition means that the signal is integrable, which is normally the case for technical signals. The Fourier transformation is defined as:

$$\mathcal{F}(f(t)) = F(\omega) = \int_{-\infty}^{\infty} f(t) \cdot e^{-j\omega t} dt \tag{2.2}$$

with $\omega = 2\pi f$, where f is the frequency. The inverse Fourier transformation is defined as:

$$\mathcal{F}^{-1}(F(\omega)) = f(t) = \frac{1}{2\pi} \int_{-\infty}^{\infty} F(\omega) \cdot e^{j\omega t} dt. \tag{2.3}$$

The power spectral density (PSD) of a signal can be obtained by the absolute value of the squared spectrum. The unit of the PSD is W/Hz. For UWB applications, dBm/MHz is often used where 0 dBm equals 1 mW. The representation in terms of PSD is used, for example, to check if the signal fits into the allocated frequency mask.

There are measures to describe the signal in the TD or in the FD. One example in the FD is the so-called group delay versus frequency, which is

$$\tau_g = -\frac{1}{2\pi} \cdot \frac{d\varphi(f)}{df}, \tag{2.4}$$

where $\varphi(f)$ is the phase of the spectrum and f is the frequency of the signal. A constant group delay means linear phase behavior, which is often required. Examples of measures of a signal in the TD are the pulse width and the pulse repetition time.

Besides UWB signals, UWB components (such as antennas, filters) and the UWB propagation channel can be described in the TD and the FD respectively. In the FD, the behavior is characterized by a complex transfer function (amplitude and phase information). An inverse Fourier transformation of the FD signal leads to the impulse response in the TD. For example, a function that is constant versus $f \in \Re$ in the FD (= flat spectrum) corresponds to an impulse at $t = 0$ in the TD. Band limitation in the FD by any system component can cause spreading of the impulse and lead to signal degradation. Detailed definitions on the measures in the TD and FD are provided in Section 2.3. Further information can also be found in books on UWB fundamentals [120, 141].

In the beginning many new and exciting topics arose from the exploration of UWB time domain short pulse modulation schemes that complemented the known frequency domain sinusoidal carrier-based systems. However the range of modulation solutions is wider and the wideband frequency domain modulation schemes showed their advantages. In between time domain and frequency domain, solutions like direct sequence spread spectrum modulation schemes and wideband carrier-based OFDM signals also exist, which look like passband pulses in the time domain. Depending on the modulation scheme and application, the RF components must be characterized in the TD, FD, or even both. To check if a signal is compliant with the allocated spectral mask, a representation in the FD is always necessary.

2.1.2 UWB channel in the frequency domain

For the FD description it is assumed that the transmit antenna is excited with a continuous wave signal with frequency f. The relevant parameters for the FD link description are:

- amplitude of transmit signal $U_{\text{Tx}}(f)$ (V)
- amplitude of receive signal $U_{\text{Rx}}(f)$ (V)
- radiated field strength $\mathbf{E}_{\text{Tx}}(f, r, \theta_{\text{Tx}}, \psi_{\text{Tx}})$ (V/m) at distance r from antenna
- transfer function of the transmit antenna $\mathbf{H}_{\text{Tx}}(f, \theta_{\text{Tx}}, \psi_{\text{Tx}})$ (m)
- transfer function of the receive antenna $\mathbf{H}_{\text{Rx}}(f, \theta_{\text{Rx}}, \psi_{\text{Rx}})$ (m)
- characteristic transmit antenna impedance $Z_{\text{C,Tx}}(f)$ (Ω)
- characteristic receive antenna impedance $Z_{\text{C,Rx}}(f)$ (Ω)
- antenna gain $G(f, \theta, \psi)$
- distance between Tx–Rx antennas r_{TxRx} (m).

The antenna transfer function is a two-dimensional vector with two orthogonal polarization components. The unit of the transfer function is meter and it is equivalent to an effective antenna height, depending on frequency [157]. The characteristic antenna impedance defines the air interface reflection coefficient. $\mathbf{H}_{\text{Tx}}(f, \theta_{\text{Tx}}, \psi_{\text{Tx}})$ is the transfer function of the transmit antenna that relates the transmit signal $U_{\text{Tx}}(f)$ to the radiated field strength $\mathbf{E}_{\text{Tx}}(f, r)$ at a distance r for an antenna in the transmit mode:

$$\frac{\mathbf{E}_{\text{Tx}}(f, r)}{\sqrt{Z_0}} = \frac{e^{-j\omega r/c_0}}{2\pi r c_0} \mathbf{H}_{\text{Tx}}(f, \theta_{\text{Tx}}, \psi_{\text{Tx}}) \cdot j\omega \frac{U_{\text{Tx}}(f)}{\sqrt{Z_{\text{C,Tx}}}}. \quad (2.5)$$

With the transfer function of the receive antenna $\mathbf{H}_{\text{Tx}}(f, \theta_{\text{Tx}}, \psi_{\text{Tx}})$ the received signal amplitude $U_{\text{Rx}}(f)$ can be related to the incident field $\mathbf{E}_{\text{Rx}}(f, \mathbf{r})$ (in the frequency domain)

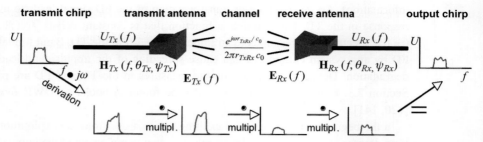

Figure 2.2 Frequency domain system link level characterization for free space. ©2009 IEEE; reprinted with permission from [179].

at an antenna in the receive mode:

$$\frac{U_{\text{Rx}}(f)}{\sqrt{Z_{\text{C,Rx}}}} = \mathbf{H}_{\text{Rx}}^T(f, \theta_{\text{Rx}}, \psi_{\text{Rx}}) \cdot \frac{\mathbf{E}_{\text{Tx}}(f, r_{\text{TxRx}})}{\sqrt{Z_0}}, \tag{2.6}$$

with $\mathbf{H}_{\text{Rx}}^T$ being the transpose of $\mathbf{H}_{\text{Rx}}$. The total analytical description of a LOS free-space UWB propagation link is given by:

$$\frac{U_{\text{Rx}}(f)}{\sqrt{Z_{\text{C,Rx}}}} = \mathbf{H}_{\text{Rx}}^T(f, \theta_{\text{Rx}}, \psi_{\text{Rx}}) \cdot \frac{e^{-j\omega r_{\text{TxRx}}/c_0}}{2\pi r_{\text{TxRx}} c_0} \cdot \mathbf{H}_{\text{Tx}}(f, \theta_{\text{Tx}}, \psi_{\text{Tx}}) \cdot j\omega \frac{U_{\text{Tx}}(f)}{\sqrt{Z_{\text{C,Tx}}}}. \tag{2.7}$$

With these parameters, the Tx–Rx free-space UWB link is illustrated in Fig. 2.2. In the frequency domain description the consecutive subsystem parameters are multiplied. The small graphs symbolize the typical influence of the link contributions. The initial chirp and its derivatives are sketched. Since antennas are reciprocal, so are their transfer functions. Therefore the transfer function of an antenna $\mathbf{H}_{\text{ant}}$ can be used at both ends of the channel; however, the direction of the signal flow with respect to the coordinate system has to be taken into account (see transposed $\mathbf{H}_{\text{Rx}}$ in (2.6)). Two orthogonal polarizations are included in the Tx and Rx transfer functions, as noted above. While in narrow band systems the radiation angles θ and ψ influence only the polarization, amplitude and the phase of the signal, in UWB systems they also influence the entire frequency-dependent signal characteristics.

For UWB links in rich scattering environments (e.g. indoors), the influence of the multipath propagation must be added to (2.7). The multipath radio propagation channel can be described by a frequency-dependent full polarimetric channel transfer matrix $\mathbf{H}_{\text{PC}}(f, \theta_{\text{Tx}}, \psi_{\text{Tx}}, \theta_{\text{Rx}}, \psi_{\text{Rx}})$. The total analytical description of a multipath UWB propagation link can then be given by

$$\frac{U_{\text{Rx}}(f)}{\sqrt{Z_{\text{C,Rx}}}} = \int_{\theta_{\text{Tx}}=0}^{\pi} \int_{\psi_{\text{Tx}}=0}^{2\pi} \int_{\theta_{\text{Rx}}=0}^{\pi} \int_{\psi_{\text{Rx}}=0}^{2\pi} \left[\mathbf{H}_{\text{Rx}}^T(f, \theta_{\text{Rx}}, \psi_{\text{Rx}}) \right. $$

$$\left. \cdot \mathbf{H}_{\text{PC}}(f, \theta_{\text{Tx}}, \psi_{\text{Tx}}, \theta_{\text{Rx}}, \psi_{\text{Rx}}) \cdot \mathbf{H}_{\text{Tx}}(f, \theta_{\text{Tx}}, \psi_{\text{Tx}}) \right] \cdot j\omega \frac{U_{\text{Tx}}(f)}{\sqrt{Z_{\text{C,Tx}}}}. \tag{2.8}$$

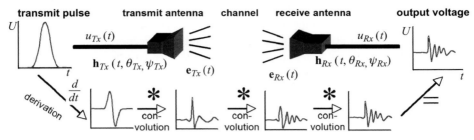

Figure 2.3 UWB system link level characterization in the time domain for free space. ©2009 IEEE; reprinted with permission from [179].

2.1.3 UWB channel in the time domain

For the time domain description it is assumed that the transmit antenna is excited with a Dirac pulse. The elements of the UWB time domain link characterization are:

- amplitude of transmit signal $u_{\mathrm{Tx}}(t)$ (V)
- amplitude of receive signal $u_{\mathrm{Rx}}(t)$ (V)
- impulse response of the transmit antenna $\mathbf{h}_{\mathrm{Tx}}(t, \theta_{\mathrm{Tx}}, \psi_{\mathrm{Tx}})$ (m/s)
- impulse response of the receive antenna $\mathbf{h}_{\mathrm{Rx}}(t, \theta_{\mathrm{Rx}}, \psi_{\mathrm{Rx}})$ (m/s)
- radiated field strength $\mathbf{e}(t, r, \theta_{\mathrm{Tx}}, \psi_{\mathrm{Tx}})$ (V/m) at position r
- distance between Tx–Rx antennas r_{TxRx} (m).

The antenna's transient impulse response is dependent on time, and also on the angles of departure θ_{Tx}, ψ_{Tx}, the respective angles of arrival θ_{Rx}, ψ_{Rx}, and the polarization [159]. As a consequence the antennas do not radiate the same pulse in all directions, which may cause severe problems in UWB communications and radar. For example, in the case of a multipath environment it is very important to include the angular behavior of the antennas in the system description since all transmitted or received paths are weighted by the antenna's characteristics, and therefore contribute with different time domain characteristics (e.g. polarization, amplitude, phase and delay) to the received voltage $u_{\mathrm{Rx}}(t)$. In Fig. 2.3 the free-space time domain link level scheme is shown. The small graphs symbolize the typical influence of the link contributions. The initial pulse and its derivative are sketched.

Since antennas do not radiate DC signals, any antenna will differentiate the radiated signal. Analog to (2.5) and (2.7), the LOS free-space time domain link can be given by:

$$\frac{\mathbf{e}_{\mathrm{Tx}}(t, \mathbf{r})}{\sqrt{Z_0}} = \frac{\delta(t - \frac{r}{c_0})}{2\pi r_{\mathrm{TxRx}} c_0} * \mathbf{h}_{\mathrm{Tx}}(t, \theta_{\mathrm{Tx}}, \psi_{\mathrm{Tx}}) * \frac{\partial}{\partial t} \frac{u_{\mathrm{Tx}}(t)}{\sqrt{Z_{\mathrm{C,Tx}}}}, \tag{2.9}$$

$$\frac{u_{\mathrm{Rx}}(t)}{\sqrt{Z_{\mathrm{C,Rx}}}} = \mathbf{h}_{\mathrm{Rx}}^T(t, \theta_{\mathrm{Rx}}, \psi_{\mathrm{Rx}}) * \frac{\delta(t - \frac{r_{\mathrm{TxRx}}}{c_0})}{2\pi r_{\mathrm{TxRx}} c_0} * \mathbf{h}_{\mathrm{Tx}}(t, \theta_{\mathrm{Tx}}, \psi_{\mathrm{Tx}}) * \frac{\partial}{\partial t} \frac{u_{\mathrm{Tx}}(t)}{\sqrt{Z_{\mathrm{C,Tx}}}}. \tag{2.10}$$

The fundamental multiplication operation in the FD corresponds to a convolution in the TD. Equation (2.9) relates the radiated field strength $\mathbf{e}_{\mathrm{Tx}}(t, r)$ at the distance r to the excitation voltage $u_{\mathrm{Tx}}(t)$ and the transient response of the transmit antenna $\mathbf{h}_{\mathrm{Tx}}(t, \theta_{\mathrm{Tx}}, \psi_{\mathrm{Tx}})$

[45]. In (2.10) again only free space LOS propagation is regarded (line of sight between Tx and Rx). Also the antenna's transient response function h_{ant} is reciprocal, so it can be applied at either Tx or Rx, but again the direction of signal flow with respect to the coordinate system has to be taken into account. The antennas are an essential part of any wireless system and their properties have to be considered carefully during all steps of the system design. For UWB impulse systems this is vital. For rich scattering environments (2.10) can be extended analog to (2.8).

2.2   UWB propagation channel modeling

To calculate exactly the wave propagation between two antennas in a given scenario, one would have to solve the Maxwell equations numerically. An investigation using a finite difference time domain solution can be found in [163]. Due to the large ratio between scenario size and wavelength this is extremely time- and memory-consuming, or even impossible for nearly all interesting scenarios. Therefore an approximation is usually used for propagation channel modeling: geometrical optics [23], where each propagation path between transmitter and receiver with all its reflections, diffractions, transmissions and scattering processes is modeled by a multipath component, usually called a "ray". With N being the number of discrete multipath components, (2.8) changes to

$$\frac{U_{Rx}(f)}{\sqrt{Z_{C,Rx}}} = \sum_{n=1}^{N} \mathbf{H}_{Rx}^{T}(f, \theta_{Rx,n}, \psi_{Rx,n}) \cdot \mathbf{H}_{PC,n}(f) e^{j\omega r_{TxRx,n}/c_0}$$
$$\cdot \mathbf{H}_{Tx}(f, \theta_{Tx,n}, \psi_{Tx,n}) \cdot j\omega \frac{U_{Tx}(f)}{\sqrt{Z_{C,Tx}}}$$

(2.11)

with

- number of multipath components N and multipath component index n
- total path length of multipath components $r_{TxRx,n}$
- transmit direction of multipath component given by $\theta_{Tx,n}$ and $\psi_{Tx,n}$
- receive direction of multipath component given by $\theta_{Rx,n}$ and $\psi_{Rx,n}$
- frequency-dependent full polarimetric channel transfer matrix of nth multipath component $\mathbf{H}_{PC,n}(f)$.

In scenarios with no other relevant multipath component other than just a line-of-sight (LOS) path, the attenuation of a signal can be determined by the free-space attenuation of the single LOS path only (see (2.7)). Assuming isotropic antennas, the free-space attenuation $L_{FS}(f)$ at a frequency f can be described by the Friis equation

$$L_{FS}(f) = \left(\frac{\lambda}{4\pi r_{TxRx}}\right)^2 \sim \frac{1}{f^2}$$

(2.12)

where r_{TxRx} denotes the distance between Tx and Rx. The UWB free-space propagation – the total attenuation between a lower and an upper frequency f_l and f_u, respectively – is

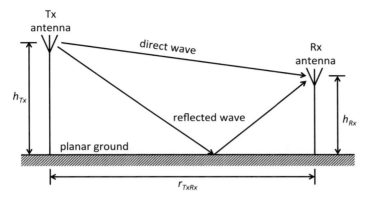

Figure 2.4 Two-path model.

approximated by an extended Friis equation according to [138]:

$$L_{\text{FS,UWB}} = \frac{c_0^2}{(4\pi\, r_{\text{TxRx}})^2 f_{\text{l}} f_{\text{u}}} \cdot \tag{2.13}$$

In cases where the LOS path is attenuated or blocked and other multipath components have a considerable field strength, more sophisticated propagation channel models are required. The most simple propagation models for general scenarios only describe the pathloss between transmitter and receiver, empirically based on the LOS path according to (2.7) and some additional parameters which have been extracted from measurements [108]. These models are the least accurate. They neither consider the actual environment nor model the individual delays of the multipath components. Particularly in the UWB case, the latter becomes very important since a high signal bandwidth results in a good time resolution. Therefore already small differences in the delays of different propagation paths may cause an effect on the UWB signal.

More sophisticated than empirical models are stochastic and deterministic propagation models. Stochastic propagation models allow the production of realistic propagation channels, but not for a specific environment; examples can be found in [109, 194]. Deterministic propagation models however first try to find all relevant propagation paths between transmitter and receiver and then calculate their contribution to the channel by geometrical optics. An example of a hybrid deterministic-stochastic UWB propagation model can be found in [72]. In the following, the special case of the deterministic propagation model is discussed: the two-path model. After that, the general ray-tracing approach is presented.

2.2.1   Two-path model

At a given frequency, a channel can also be modeled by a so-called two-path model, where the first path is the direct path between transmitter and receiver and the second path typically results from a ground reflection (see Fig. 2.4). Taking into account parameters such as the ground reflection coefficient $\gamma\, e^{j\phi}$, the transmitter height h_{Tx} and the receiver

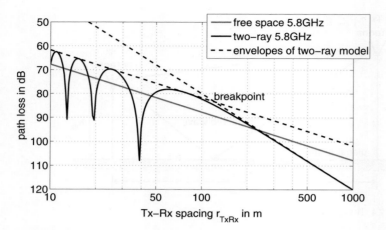

Figure 2.5 Path loss for narrowband two-path model ($h_{\mathrm{Tx}} = 1$ m, $h_{\mathrm{Rx}} = 1$ m, perfect conducting ground: $\gamma = 1, \phi = \pi$).

height h_{Rx}, the path loss L in dB at one frequency as a function of distance is described in [55, 148] as

$$L = -10 \log \left[\left(\frac{\lambda}{4\pi r_{\mathrm{TxRx}}} \right)^2 \left\{ 1 + \gamma^2 + 2\gamma \cos \left(\frac{2\pi \Delta l}{\lambda} + \phi \right) \right\} \right] \qquad (2.14)$$

with Δl being the difference in path length between the direct wave and the reflected wave. In Fig. 2.5 the path loss of the two-path model is compared to free space at a frequency of 5.8 GHz. Due to the superposition of the two waves, the path loss as a function of the distance shows deep fading at distances below the breakpoint where Δl is below $\lambda/2$. The figure shows that the envelope of the path loss increases at 20 dB per decade for small distances and 40 dB per decade for large distances above the breakpoint, due to the interference of the two waves. Channels that can be approximated by the two-path model may hence present locations with a very low receive power when operating the system with a narrow bandwidth.

After considering the two-path model at a given frequency, its extension to the UWB case is presented. It is shown in [148] that the UWB signal attenuation for a signal with contributions between f_{l} and f_{u} can be estimated by

$$L_{\mathrm{UWB}} = -10 \log \left[\frac{1}{f_{\mathrm{u}} - f_{\mathrm{l}}} \int_{f_{\mathrm{l}}}^{f_{\mathrm{u}}} \left(\frac{c_0}{4\pi f r_{\mathrm{TxRx}}} \right)^2 \right.$$

$$\left. \cdot \left\{ 1 + \gamma^2 + 2\gamma \cos \left(\frac{2\pi f \Delta l}{c_0} + \phi \right) \right\} df \right]. \qquad (2.15)$$

If $r_{\mathrm{TxRx}} \gg h_{\mathrm{Tx}}, h_{\mathrm{Rx}}$ then

$$\Delta l \cong \frac{2 h_{\mathrm{Tx}} h_{\mathrm{Rx}}}{r_{\mathrm{TxRx}}} \qquad (2.16)$$

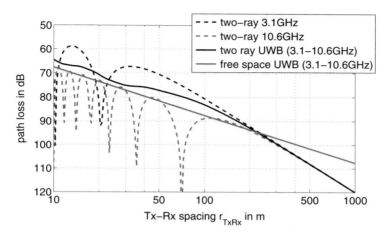

Figure 2.6 Path loss for UWB two-path model ($h_\mathrm{Tx} = 1$ m, $h_\mathrm{Rx} = 1$ m, perfect conducting ground: $\gamma = 1, \phi = \pi$).

applies. An exemplary behavior of the UWB two-path model versus distance is shown in Fig. 2.6. The figure shows that there are no longer any deep fading holes when compared to the behavior of the narrowband two-path model.

2.2.2   Ray-tracing for UWB channel modeling

The channel models described so far do not consider the specific environment of a typical wireless scenario, where radio waves propagate on various paths from transmitter to receiver. More realistic deterministic models also consider the effect of multipath propagation as a result of reflection, transmission, scattering and diffraction of waves in a given scenario. This leads to 2- or even 3-dimensional channel models that describe the propagation at a given frequency. Ultra-wideband propagation can be modeled by subdividing the ultra-wide bandwidth into a set of frequencies. For all relevant multipath components n from (2.11), the full polarimetric channel transfer matrix $\mathbf{H}_\mathrm{PC,n}(f)$ is then determined by the channel model, depending on the frequency f based on geometrical optics as explained before. The geometrical traces of each multipath component are determined by a method called ray-tracing, which is a popular method in computer graphics as well. More information can be found in [51, 55].

To visualize effects of the channel and the antennas in the UWB case, simulations based on a ray-tracing model are given in the following. Figure 2.7 shows a typical indoor scenario. The polygon model describes a laboratory scenario including furniture (tables, cupboard) and iron objects (instruments, table legs). The physical properties of the objects are modeled by their complex permittivity ϵ, permeability μ and the standard deviation of the surface roughness σ. Table 2.1 summarizes the physical parameters of the objects. To describe the indoor channel for an ultra-wide bandwidth, the complex transmission coefficient is determined for a large set of frequencies from 2.5 to

Table 2.1 Properties of the indoor channel objects.

Object	ε_r	μ_r	σ_r mm
Walls	$5 - j0.1$	1	1
Instruments, table legs	$1 - j10^9$	10	0.01
Furniture	$2.5 - j0.1$	1	0

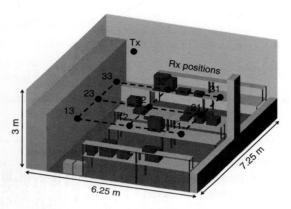

Figure 2.7 Indoor laboratory scenario. ©2010 KIT Scientific Publishing; reprinted with permission from [168].

12.5 GHz with a frequency step width of 6.25 MHz. This allows the determination of the frequency-dependent UWB propagation link according to (2.11). The resulting channel transfer function, including antenna effects, is then integrated into the system simulator. A detailed description of the ray-tracing principle used here is available in [51]. In Fig. 2.7, one Tx and nine different LOS Rx positions (11–33) are defined. The height of both the transmitter and the receiver is 2 m. The antennas are oriented as shown in Fig. 2.8 with their ground planes parallel to the floor of the room. To study the effects of the channel and the antennas separately, two different cases are investigated.

First, we will look at the antenna effects; so all multipaths (except for the LOS path) are turned off, and the antenna models are activated. The transmission behavior of the resulting free-space UWB channel, including antenna effects, is shown in Fig. 2.9 (left). In theory, the free-space attenuation according to (2.12) results in an attenuation of 20 dB per frequency decade. This principal behavior can also be seen in Fig. 2.9 (left). However, the imperfections of the realized antenna characteristics lead to some additional frequency-dependent variations. The associated group delay τ_g is shown in Fig. 2.9 (right). It can be seen that the antennas cause a group delay variation of about 1 ns inside the relevant frequency range due to manufacturing tolerances. The mean value of the group delay is related to the physical distance between Tx and Rx, and to delays introduced by the analog filter and the antennas.

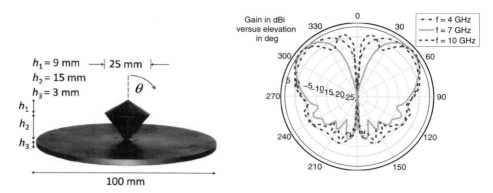

Figure 2.8 Properties of the Monocone antenna. Left: ©2009 De Gruyter; reprinted with permission from [169]; right: ©2010 KIT Scientific Publishing; reprinted with permission from [168].

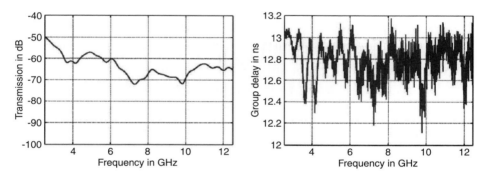

Figure 2.9 Left: free-space channel including antenna effects in the frequency domain; right: associated group delay. ©2010 KIT Scientific Publishing; reprinted with permission from [168].

In a second step, the multipath contributions found by the ray-tracing method are turned on. Antenna effects remain active. The transmission behavior of the multipath UWB channel, including antenna effects, is visualized in Fig. 2.10 (left). The exemplary group delay behavior can be seen in Fig. 2.10 (right). The comparison between Figs. 2.9 (left) and 2.10 (left) shows that the multipaths lead to a further distortion of the attenuation profile. At several equally spaced frequencies, the attenuation increases strongly. This is a clear sign for the existence of a second strong propagation path in addition to the LOS path. This is the same behavior as described by the two-path model. A comparison of the group delay behavior between Figs. 2.9 (right) and 2.10 (right) shows that the multipath channel leads to a large group delay variation, which is of the order of 10 ns. This variation is 10 times larger than the group delay variation of the analog filter which again points out why the antenna behavior is very critical in UWB systems. The mean value of the group delay differs slightly between Figs. 2.9 (right) and 2.10 (right) due to a different Tx–Rx distance.

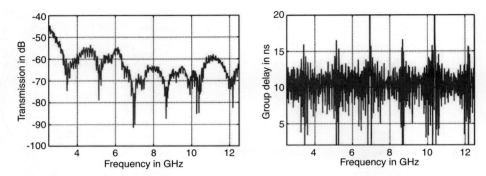

Figure 2.10 Left: multipath channel including antenna effects in the frequency domain; right: associated group delay. ©2010 KIT Scientific Publishing; reprinted with permission from [168].

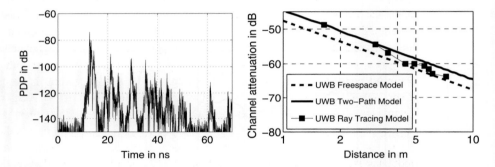

Figure 2.11 Left: power delay profile of the multipath channel, including antenna effects; right: channel attenuation vs distance for different channel models. ©2010 KIT Scientific Publishing; reprinted with permission from [168].

Figure 2.11 (left) visualizes the properties of the UWB multipath channel including antenna effects in terms of a power delay profile (PDP). The figure shows the distribution and the power of the individual paths versus arrival time. The figure is given for a constellation with a Tx–Rx distance of 3.57 m, which corresponds to a 11.9 ns delay for the LOS path. Since the analog filter and the physical antennas add further delays of the order of 1 ns in total, the strongest path in the figure occurs after about 13 ns. In Fig. 2.11 (right), all ray-tracing results are plotted over distance in comparison to the UWB free-space model and the UWB two-path model. Hence the investigated LOS ray-tracing channels are simulated for the Rx positions from Fig. 2.7, by neglecting the antenna influence so that only the channel behavior is compared. The parameters of the UWB two-path channel model are $\gamma e^{j\phi} = -1$ for a perfect electric conducting surface representing the strongest possible reflection, and $h_{Tx} = h_{Rx} = 2$ m, which corresponds to the heights of the ray-tracing scenario. Figure 2.11 (right) shows that the attenuation of the ray-tracing UWB channel model is in between the well-known free-space model and the two-path model for metallic reflection. This makes sense, since the physical reflection coefficient of a second path may not always be 1 (absolute value), but somewhere in between 0 and 1. This means that the actual UWB attenuation in LOS scenarios with only two strong paths ranges between the bounds of the free-space model and the two-path model for metallic reflection. Principal fading characteristics are presented in the next section. Detailed verification of the ray-tracing based UWB channel model (by comparing simulation and measurement data) shown here, can be found in [135]. Furthermore, [135] verifies an extended FDTD/ray-tracing model, which is not applied here because of its complexity.

2.2.3   Fading

Multipath propagation leads to a strongly varying received power. This effect is called fading and occurs when signals at a given frequency interfere with each other, as will happen when the same signal arrives at the receiver from different directions due to multipath propagation. Depending on the relative phase between the signals, they can cancel each other out, as has been shown before for the two-path case. This leads to so-called fading holes in which a receiver may no longer be able to demodulate a signal due to insufficient signal-to-noise ratio. Spatial fading means that the signal power varies strongly around a given position. The fading characteristics may also vary with time due to time-dependent varying propagation conditions. In general, fading is an unwanted effect. When the bandwidth of the signal increases, the fading of the total received power becomes weaker – since the locations of the fading holes vary with frequency. In other words, the phase of the signals varies with frequency, so at some frequencies the signals cancel each other out, at other frequencies they accumulate. Averaging over a wide frequency range mitigates the fading. Figure 2.12 visualizes the total received power in an indoor environment for two cases (top: WLAN signal with 20 MHz bandwidth; bottom: UWB signal with 7.5 GHz bandwidth). It can be seen that the power distribution is much more homogeneous for the UWB signal. The fact that UWB signals are much less affected by fading is one of the important advantages of UWB over conventional narrowband technologies.

2.3   Parameters for UWB RF system and component characterization

In Fig. 2.13 the functional blocks of a typical UWB radio system are given. In UWB systems not only must the radio channel be considered as frequency dependent, but all other components as well. In contrast to classic narrowband RF system theory, where the characteristics of the RF system and its blocks are regarded for only a small bandwidth and are typically considered to be frequency independent, the characterization of RF system blocks over an ultra-wide frequency range requires new specific quantities and representations (for antennas see [149, 154]).

As can be seen directly from (2.8) antennas are, along with the channel, the most critical components from that perspective in the system. An impulse fed to a UWB antenna is subject to differentiation, dispersion, radiation and losses (dielectric/ ohmic). Besides the well know radiation pattern, one has to carefully consider that

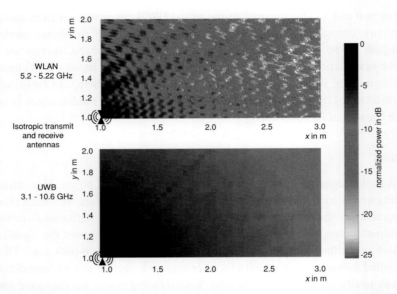

Figure 2.12 Power distribution of UWB signal vs WLAN signal in a lossy room based on ray-tracing simulations.

Figure 2.13 Functional blocks of a radio system.

the antenna's frequency dependency might vary with transmit or receive angle. Therefore, in the following, the UWB system parameters are mainly explained for antennas, but all parameters given in the following can be applied to all other system components such as amplifiers and filters as well. Due to the reciprocity of antennas, the indices Tx and Rx are omitted. In this section both time domain and frequency domain representations are regarded. Depending on the application, the relevant ones have to be selected. In general, the Fourier transforms forward and backward are the operations to switch from the frequency domain to the time domain and vice versa.

The channel's or antenna's complete behavior, including frequency dependency, can be described by the linear system theory. The characteristics are either expressed by a time domain impulse response $\mathbf{h}(t, \theta_{Tx}, \psi_{Tx})$ or the frequency domain transfer function $\mathbf{H}(f, \theta_{Tx}, \psi_{Tx})$, as given earlier, both of which contain the full information on the antenna radiation or reception, respectively. The dispersion of the antenna can be analyzed by regarding the analytic impulse response, which is calculated by the Hilbert transform $\mathcal{H}$ commonly used in signal processing:

$$h^+(t) = (h(t) + j\mathcal{H}\{h(t)\}). \tag{2.17}$$

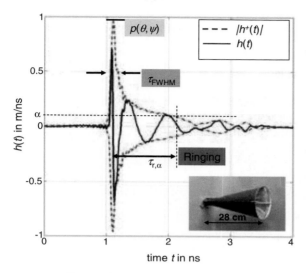

Figure 2.14 Characterization of the antenna via its time domain transient response (in this case, horn antenna). ©2009 IEEE; reprinted with permission from [179].

The envelope $|h^+(t)|$ of the analytic impulse response localizes the distribution of energy versus time and is a direct measure for the dispersion of an antenna. A typical example of a measured antenna impulse response $h(t)$ is shown together with $|h^+(t)|$ in Fig. 2.14 for a given polarization and direction (θ, ψ) of radiation. A typical example of a squared channel impulse response (power delay profile) is shown in Fig. 2.11 (left). In the following, the most important TD and FD parameters are introduced. Note that all parameters for antennas are dependent on polarization and spatial coordinates (r, θ, ψ). Examples for specific antennas are given later in the book.

2.3.1   Delay spread of the radio channel

The parameter delay spread τ_{DS} measures the multipath behavior of the radio channel. It is the second moment of the power delay profile (absolute square of channel impulse response) given by

$$\tau_{DS} = \sqrt{\frac{\int_0^\infty (\tau - \overline{\tau}_D)^2 |h^+(\tau)|^2 d\tau}{\int_0^\infty |h^+(\tau)|^2 d\tau}} \qquad (2.18)$$

with the average delay time

$$\overline{\tau}_D = \frac{\int_0^\infty \tau |h^+(\tau)|^2 d\tau}{\int_0^\infty |h^+(\tau)|^2 d\tau} . \qquad (2.19)$$

Principally, one could also use the previously defined delay spread to characterize the dispersion behavior of all the other system components besides the propagation channel, but for antennas and other frontend components a different set of parameters is usually defined (see sections on envelope width and ringing).

2.3.2   Peak value of the envelope

The peak value $p(\theta, \psi)$ of the analytic envelope $|h^+(t)|$ is a measure for the maximum value of the strongest peak of the antenna's time domain transient response envelope (see Fig. 2.14). It is mathematically defined as

$$p(\theta, \psi) = \max_t |h^+(t, \theta, \psi)| \tag{2.20}$$

and has the unit m/s. A high peak value $p(\theta, \psi)$ is desirable.

2.3.3   Envelope width

The envelope width describes the broadening of the radiated impulse and is defined as width of the magnitude of the analytic envelope $|h^+(t)|$ at half maximum (FWHM). Analytically it is defined as:

$$\tau_{\text{FWHM}} = t_2 \,|_{|h^+(t_1)|=p/2} - t_1 \,|_{t_1 < t_2, |h^+(t_2)|=p/2}. \tag{2.21}$$

The envelope width should not exceed a certain value (typically a few hundred picoseconds for FCC UWB systems) in order to ensure high data rates in communications or high resolution in radar applications.

2.3.4   Ringing

The ringing of a UWB antenna is undesirable and is usually caused by resonances due to energy storage or multiple reflections in the antenna. It results in oscillations of the radiated pulse after the main peak. The duration of the ringing τ_r, defined as the time until the envelope has fallen from the peak value $p(\theta, \psi)$ below a certain lower bound $\alpha \cdot p(\theta, \psi)$, is measured as follows:

$$\tau_{r,\alpha} = t_2 \,|_{h^+(t_2)=p} - t_1 \,|_{t_1 < t_2, h^+(t_1)=\alpha p}. \tag{2.22}$$

Typically, a value for α of between 0.1 and 0.25 is specified. The duration τ_r of the ringing of a UWB antenna should be negligibly small, i.e. less than a few envelope widths (τ_{FWHM}). The energy contained in ringing is of no use at all and it lowers the peak value $p(\theta, \psi)$. In a UWB antenna it can therefore be eliminated by, for example, absorbing materials.

2.3.5   Transient gain

The transient gain $g_T(\theta, \psi)$ is an integral quality measure, which in the case of a transmitting antenna characterizes the ability of an antenna to radiate the power of a given waveform $u(t)$ or $U(f)$:

$$g_T(\theta, \psi) = \frac{\|h(t, \theta, \psi) \cdot \frac{du(t)}{dt}\|^2}{\|\sqrt{\pi} c_0 u(t)\|^2} = \frac{\|H(\omega, \theta, \psi) j\omega U(f)\|^2}{\|\sqrt{\pi} c_0 U(f)\|^2} \qquad (2.23)$$

where the norm is defined by

$$\|f(x)\| = \int_{-\infty}^{\infty} |f(x)| \, dx. \qquad (2.24)$$

2.3.6   Gain in the frequency domain

The gain in the frequency domain is defined as in narrowband systems. It can be calculated from the antenna transfer function but will be frequency dependent:

$$G(f, \theta, \psi) = \frac{4\pi f^2}{c_0^2} |H(f, \theta, \psi)|^2 \qquad (2.25)$$

Note that in the case of an antenna, it is important that the transfer function is multiplied by f^2. During the radiation of pulses the whole bandwidth is covered quasi-simultaneously. Therefore a single entity is desired which characterizes the total amount of radiated power in the specific direction. Such a parameter could be the peak value of the envelope (cf. Fig. 2.14). However, in order to make a connection to a more popular entity, the mean gain $G_m(\theta, \psi)$ is defined in the specified bandwidth as

$$G_m(\theta, \psi) = \frac{1}{f_u - f_l} \int_{f_l}^{f_u} G(f, \theta, \psi) df \qquad (2.26)$$

where f_l is the lower cut-off frequency and f_u the upper cut-off frequency of the considered frequency range.

2.3.7   Group delay

As already shown in Section 2.2.2, the group delay $\tau_g(\omega)$ of an antenna plays an important role in UWB systems. It is defined in the FD as:

$$\tau_g(\omega) = -\frac{d\varphi(\omega)}{d\omega} = -\frac{d\varphi(f)}{2\pi \, df}, \qquad (2.27)$$

where $\varphi(f)$ is the frequency-dependent phase of the radiated signal. Also of interest is the mean group delay $\bar{\tau}_g$ as it is a single number for the whole UWB frequency range:

$$\bar{\tau}_g = \frac{1}{\omega_2 - \omega_1} \int_{\omega_1}^{\omega_2} \tau_g(\omega) d\omega. \qquad (2.28)$$

A nondistorted structure is characterized by a constant group delay, i.e. a linear phase in a relevant frequency range. The nonlinearities of a group delay indicate the resonant character of the device, which affects the ability of the structure to store the energy and results in ringing and oscillations of the antenna impulse response $h(t)$ [82]. A measure for the constancy of the group delay is the deviation from the mean group delay $\bar{\tau}_g$,

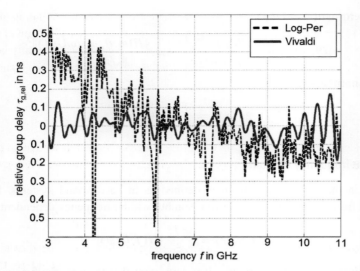

Figure 2.15 Relative group delay $\tau_{g,rel}(f)$ of a Vivaldi antenna and a log-periodic antenna. ©2009 IEEE; reprinted with permission from [179]. For details regarding the antennas, see Chapter 3.

denoted as relative group delay:

$$\tau_{g,rel}(\omega) = \tau_g(\omega) - \bar{\tau}_g. \tag{2.29}$$

Examples of the relative group delay of a Vivaldi antenna and a log-periodic antenna (see Section 3.2) in the frequency range from 3 to 11 GHz are shown in Fig. 2.15. In the case of the Vivaldi antenna, which is a nonresonant structure, the relative group delay shows only weak and slow oscillations over the whole frequency band. However the relative group delay of the log-periodic antenna shows strong and sharp oscillations over the whole frequency band, which results in an oscillation of the antenna impulse response $h_{Log-Per}(t)$. For this antenna the group delay is frequency dependent: lower frequencies show a higher relative group delay. This behavior is caused by the frequency-dependent phase center of the radiation.

2.3.8   Fidelity

From Maxwell's equations it is well known that the signals radiated by an antenna result from the distribution of the exciting currents on the antenna. In narrowband applications it can be assumed that this current distribution is constant over frequency, which results in a given, constant radiation characteristic. Also, the transmitted sinusoidal signals remain sinusoidal like the currents and can easily be detected in any direction of the radiation. For ultra-wideband time domain operation this is not usually the case. Over the wide bandwidth the current distribution changes which causes a variation of the radiated signal over the direction. In addition, the currents are differentiated for radiation (see Maxwell's equations). The signals are distorted with respect to a reference signal, e.g. the original signal or the signal in the main beam direction. This behavior has a significant impact on most time domain impulse UWB applications. Since the radiation direction is usually unknown, or in the multipath case there is not one discrete direction of radiation, the distortion cannot be compensated for in the system. The so-called fidelity is typically used as a measure for the distortion, which can be defined based on the signal voltage $u(t)$ in relation to a reference signal $u_{ref}(t)$ as [125]:

$$F = \max_\tau \int_{-\infty}^{\infty} \frac{u(t+\tau) \cdot u_{ref}(t)}{\|u(t)\|_2 \cdot \|u_{ref}(t)\|_2} dt , \qquad (2.30)$$

where $\| \ \|_2$ indicates the 2-norm. In the case of antennas the fidelity must be defined as the correlation of the radiated impulse in the direction θ, ψ with the reference impulse $h_{ref}^+(t)$:

$$F(\theta, \psi) = \max_\tau \int_{-\infty}^{\infty} \frac{h^+(t+\tau, \theta, \psi) \cdot h_{ref}^+(t)}{\|h^+(t, \theta, \psi)\|_2 \cdot \|h_{ref}^+(t)\|_2} dt . \qquad (2.31)$$

In some cases the radiated impulse in the main beam direction is used as a reference [124]. An ideal antenna with identical impulses in all directions would have the fidelity $F = 1$. In radar and communication applications a high fidelity, i.e. $F > 0.8$ is vital, otherwise the expected signal cannot be detected due to insufficient correlation with a template or reference signal. In combination with the above described peak value of radiation $p(\theta, \psi)$ (see (2.20)), the two quantities together define the quality of the received impulse signal. Systems suffer if either of these is small.

2.4 Impulse radio versus orthogonal frequency division multiplexing

In principle, there are two approaches to exploit the allocated ultra-wide bandwidth in the lower GHz frequency range. The first approach makes use of ultra-short pulses in the time domain, where the pulse duration is of the order of some hundred picoseconds to nanoseconds. This corresponds to an ultra-wideband signal in the frequency domain. As a result the modulated pulses are radiated via an ultra-wideband antenna without a carrier in the baseband. An up-conversion to a carrier frequency is not required. This method is also called Impulse Radio (IR) and will be discussed later in the book. The second approach subdivides the allocated ultra-wide bandwidth into a set of multiple broadband channels, where the center frequency of each broadband channel defines a so-called sub-carrier. Each sub-carrier is modulated simultaneously in such a way that the sub-carrier signals are orthogonal to each other. Guard bands between the channels are not required. This approach is known as OFDM transmission [144].

One major characteristic of OFDM is that its orthogonality avoids cross-talk between the broadband channels. Since large guard bands are not required, a high spectral efficiency can be achieved. In the case of interference or frequency-dependent distortions, the advantage of OFDM over IR is the ability to cancel out impacted channels so that only part of the data may be lost. In IR, the complete pulse shape is distorted so it may not be possible to correctly demodulate the received signal. Since OFDM transmission is based on orthogonal signals with dedicated sub-carriers, the frequency synchronization between transmitter and receiver has to be very accurate. Frequency offsets would distort the orthogonality and hence the performance. For example, frequency offsets occur in the presence of a moving transmitter or receiver due to the Doppler shift. In combination with multipath propagation this may limit the performance of OFDM transmission at a high speed. Orthogonality can be implemented by using an FFT algorithm at the receiver side and an IFFT algorithm at the transmitter side. This requires increased signal processing efforts and consequently a relatively high power consumption. This is a drawback of OFDM when used for UWB communications with limited power resources (e.g. a battery in a mobile application). It has to be kept in mind that the total transmit power has to stay below approximately 0.5 mW anyway, assuming the FCC mask. Therefore it is also particularly desirable to exploit the potential of low power consumption on the baseband and signal processing side.

To summarize, the advantages of IR are (i) a simple architecture, and (ii) the ability to generate basic pulse shapes by simple analog components. A disadvantage is the limited exploitation of the bandwidth, which may however be improved by analog or digital pulse shaping circuits. OFDM presents a very good exploitation of the bandwidth. However, this requires an increased signal processing effort and an increased power consumption, which may not be compatible with mobile applications based on limited power resources. Last but not least it has to be mentioned that there is a further allocation of an ultra-wideband frequency band in the lower terahertz range. This can be exploited by an up-conversion of an ultra-wideband signal to a suited carrier frequency [80].

2.5   UWB pulse shapes and pulse shape generation

2.5.1   Classical pulse shapes

The principle of IR is to transmit the signal directly in the baseband. Since there is no carrier, the pulse shape has to cover a bandwidth of several GHz. The radiated signal has to fulfill the spectral regulation, and it is preferable to fully exploit it to achieve the maximum radiated power possible. The efficiency η of a UWB signal w.r.t. to a given regulation is calculated by integrating the power spectral density (PSD) in the relevant frequency range and by dividing this term by the maximum allocated power. Considering the FCC mask, this means

$$\eta = \frac{\int\limits_{3.1 \text{ GHz}}^{10.6 \text{ GHz}} \text{PSD}(f)\,df}{\int\limits_{3.1 \text{ GHz}}^{10.6 \text{ GHz}} \text{PSD}_{\text{Reg}}(f)\,df} = \frac{\int\limits_{3.1 \text{ GHz}}^{10.6 \text{ GHz}} \text{PSD}(f)\,df}{10^{\frac{-41.3}{10}}\,\frac{\text{mW}}{\text{MHz}}\,7.5 \text{ GHz}} = \frac{\int\limits_{3.1 \text{ GHz}}^{10.6 \text{ GHz}} \text{PSD}(f)\,df}{0.555 \text{ mW}}. \quad (2.32)$$

This efficiency is also called normalized effective signal power (NESP). Since UWB signals are very short in the time domain, the goal is to generate pulses characterized by an extremely fast rise and fall time. This can be achieved by analog circuits that make use of nonlinear components such as avalanche transistors, step recovery diodes, tunnel diodes, and nonlinear transmission lines [141, 143]. A fully integrated pulse shape generator in CMOS technology can be found in [19].

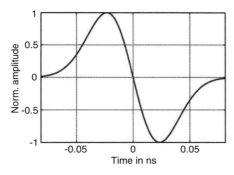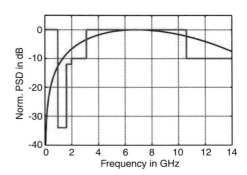

Figure 2.16 Gaussian monocycle in the time domain and in the frequency domain together with the FCC regulation. ©2010 KIT Scientific Publishing; reprinted with permission from [168].

The following paragraph considers two nonlinear components for the pulse shape generation in more detail.

- *Avalanche transistors* make use of the avalanche effect. Carriers are accelerated by an external voltage, hit valence electrons and remove them from their binding. The valence electrons are raised into the conducting band, which means that there are remaining electron-hole pairs. The accelerated carriers do not recombine but stay in the conducting band and hit further valence electrons that are raised into the conducting band. The number of carriers in the conducting band increases rapidly (like an avalanche). This effect can be used to achieve a very short rise time of a pulse [107].
- *Step Recovery Diodes* (SRDs) are doped such that the lifetime of the minority carriers is larger compared to normal diodes. This leads to storage of carriers. When the voltage is inverted at the SRD from positive to negative, the diode is conductive in both directions for a very short time. Hence the output voltage is negative. When all the charges are removed, the diode will directly block (creating a blocking zone), which means that the output voltage rises to zero. These mechanisms lead to a very short pulse at the output. The pulse width and the sign of the pulse can be controlled by a shorted transmission line [40, 81, 141].

The advantages of analog pulse generation are simplicity and cost-efficiency [180]. Pulses that are generated by this principle look like the Gaussian function or one of its derivatives [43]. Since the Fourier transformation of a Gaussian shape in the time domain results in a Gaussian shape in the frequency domain (and hence is not flat), this means that the spectral mask is not optimally exploited. Figure 2.16 shows the first derivative of the Gaussian pulse (also known as the Gaussian monocycle) in the time domain and the power spectral density in the frequency domain with the center frequency at 6.85 GHz, which is the center of the usable band in the FCC regulation. It can be seen in the figure that besides the inefficient use of the spectral mask, the mask is also violated. To avoid this violation, there are several possibilities, as follows.

- It is possible to reduce the power until the complete spectrum is below the mask. One would have to attenuate the Gaussian monocycle by about 25 dB, which decreases the efficiency to an unacceptable level.

- The Gaussian monocycle is used together with a Tx filter before the Tx antenna. The filter is designed to ensure sufficient attenuation outside the frequency range 3.1–10.6 GHz so that the mask is not violated. The efficiency of the radiated pulse is moderate, and the pulse shape after the filter is no longer the original Gaussian monocycle. In principle one could design the optimal filter response in the FD to fit the emission mask well but one has to carefully investigate its effect on the time domain behavior (e.g. with a larger envelope width and higher ringing).
- A pulse shape is generated that corresponds to a higher derivative of a Gaussian pulse. It can be shown that the fifth (or higher) derivative of a Gaussian pulse fits into the FCC mask quite well and therefore does not require an additional filter. The design of a pulse shape generator, which generates the fifth derivative of a Gaussian pulse is shown in [30, 31]. However, the efficiency of the pulse is worse compared to the filtered Gaussian monocycle. An approximation of the sixth derivative of a Gaussian pulse (six zero-crossings) is shown in Fig. 2.17. The figure also shows sampled values with a sampling time of 17.86 ps. The resulting pulse shows an efficiency η of 43.3% in the relevant frequency range 3.1–10.6 GHz.

2.5.2   Optimal pulse shapes

The efficiency can be improved by combinations of Gaussian pulses of different orders (derivatives) [191]. This requires an analog generation of several pulses in a suitable combination, which leads to increased complexity and costs. Another possibility is to generate optimal pulse shapes in a digital way and to convert these into the analog domain using a digital–analog converter (DAC) before radiating the signal via the Tx antenna. From the mathematical point of view, this requires only a transformation of the spectral mask into the time domain. However, due to the large bandwidth, the DAC must have an enormous sampling rate. This again leads to increased costs and power consumption, which contradicts the idea of UWB being a power-efficient system [177]. It is also possible to manipulate a classical pulse shape (that was generated in an analog way) by a finite impulse response (FIR) filter so that the resulting pulse is optimal in terms of efficiency. Although the optimal FIR coefficients are determined by methods of a digital filter approach, the implementation can be analog. Hence, no DAC with ultra-high speed is required. Delay lines can be programmed by a microprocessor and the basis pulse can be weighted by coefficients [17]. A resolution of picoseconds is achievable.

2.6   Modulation and coding

The following considerations are based on IR transmission. Such a UWB signal is generated by using a fixed pulse repetition time T defining the positions of the pulses before modulation and coding. However, to carry information, modulation and coding is

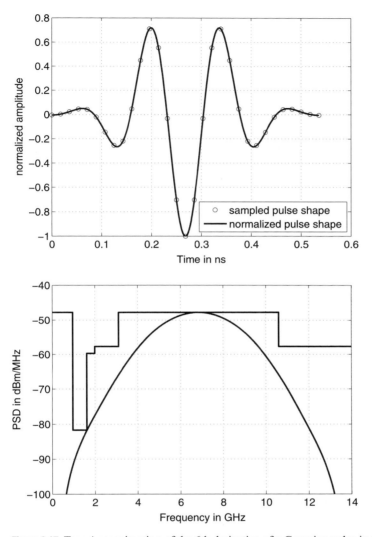

Figure 2.17 Top: Approximation of the 6th derivative of a Gaussian pulse in the time domain by multiplying a Gaussian pulse (no derivative) with a sine function; bottom: power spectral density of a pulse compared to emission mask. ©2010 KIT Scientific Publishing; reprinted with permission from [168].

required – so a bit can be represented by N_p pulses in the time domain. Important modulation techniques are pulse position modulation (PPM), pulse amplitude modulation (PAM) with the special case of on-off keying (OOK), binary phase shift keying (BPSK) and orthogonal pulse modulation (OPM). The following sections concentrate on OOK, PPM and OPM. This is motivated by the fact that PPM offers good spectral properties, OOK is easy to implement and OPM has the potential to increase the data rate.

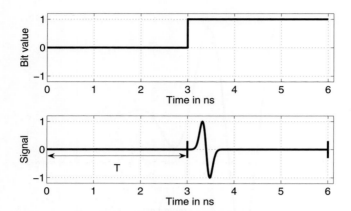

Figure 2.18 Principle of OOK modulation. ©2010 KIT Scientific Publishing; reprinted with permission from [168].

2.6.1   On–off keying

On–off keying (OOK) means that pulses are switched on (if the bit value is 1) or switched off (if the bit value is 0) at the positions defined by the pulse repetition time T to modulate the data. The principle is shown in Fig. 2.18.

2.6.2   Pulse position modulation

Pulse position modulation (PPM) means that in the time domain, pulses are either placed at the positions defined by the pulse repetition time T or at a position which has a constant time shift T_{PPM} (PPM offset) w.r.t. to this position. For example, a binary 0 is represented by a pulse without offset, while a binary 1 would introduce an offset, accordingly. Figure 2.19 visualizes the principle of PPM in the case of 1 pulse per bit, i.e. $N_{\text{p}} = 1$, where T_{p} indicates the duration of the pulse. To avoid inter-symbol interference (ISI), the PPM offset is constrained by $T_{\text{PPM}} < T$. High data rates can be achieved for a small pulse repetition time T. However, if T is so small that it is of the order of (or even smaller than) the delay spread τ_{DS} of the channel (see Section 2.3.1), inter-symbol interferences are likely to occur [182]. In addition to the delay spread of the channel, the ringing of the antennas and front end components must also be considered, although the channel usually dominates this effect. For these reasons, a lower limit has to be considered for T. In indoor channels, typical values for τ_{DS} are in the range of 10 ns [87].

The PPM offset affects the performance in terms of the bit error rate BER versus the energy per bit to noise power spectral density ratio E_{b}/N_0. To optimize the performance, T_{PPM} has to be optimized. In the case of an AWGN channel and coherent detection, the cross-correlation function between the pulses for a binary 1 and a binary 0 has to be minimized. Denoting $p(t)$ as the pulse shape without PPM offset, the cross-correlation

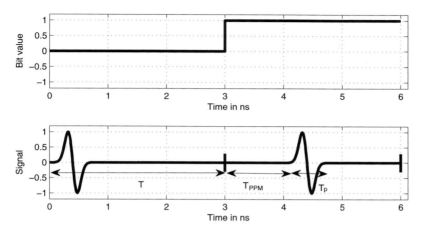

Figure 2.19 Principle of PPM. ©2010 KIT Scientific Publishing; reprinted with permission from [168].

function $r_{CCF}(T_{PPM})$ is given by

$$r_{CCF}(T_{PPM}) = \int_{-T/2}^{T/2} p(t) \cdot p(t - T_{PPM}) \, dt. \tag{2.33}$$

The minimum is obtained by

$$\frac{d}{dT_{PPM}} r_{CCF}(T_{PPM}) = 0 \tag{2.34}$$

(see [118]). Assuming a classical Gaussian monopulse, characterized by

$$p(t) = -\frac{At}{\sigma^2} e^{-\frac{t^2}{2\sigma^2}} \tag{2.35}$$

and some approximations in (2.34), lead to $T_{PPM} = 0.38985 \cdot T_p$, where T_p is the duration of the pulse [118]. A is an amplitude normalization constant, σ is related to the width of the pulse. A more accurate calculation performed in [43] leads to $0.5408 \cdot T_p$. To summarize, in AWGN channels, the optimal PPM offset is a function of the pulse shape [43].

2.6.3 Orthogonal pulse modulation

While signals with classical modulation schemes such as OOK and PPM carry only 1 bit of information per pulse (or even less, when a bit is represented by N_p pulses), orthogonal pulse modulation (OPM) carries multiple bits per pulse. This is realized by the fact that OPM makes use of a set of N_{ortho} (e.g. 2, 4, 8, 16, 32, ...) different orthogonal pulse shapes, where each pulse represents a special series of $\log_2 N_{ortho}$ (e.g. 1, 2, 3, 4, 5, ...) bits. The different pulse shapes are set at the positions defined by the pulse repetition time T. Figure 2.20 shows the principle of OPM for $N_p = 4$ pulses. The figure also indicates the pulse repetition time T. For OPM T has to be larger than or equal to the

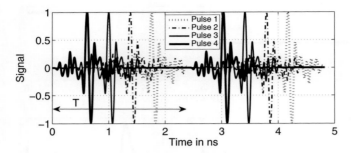

Figure 2.20 Principle of OPM. ©2010 KIT Scientific Publishing; reprinted with permission from [168].

pulse duration. In Fig. 2.20, T is identical to the pulse duration which maximizes the data rate. Compared to OOK and PPM, the data rate increases by the factor $\log_2 N_{\text{ortho}}$. Proper demodulation at the receiver side also requires the received pulse shapes to be orthogonal to each other: integrating the product of two different pulse shapes over the duration T results in zero, allowing for proper bit reconstruction. In reality, the orthogonality may suffer from the imperfections of the channel and the hardware.

2.7   Time hopping

Without modulation, an ultra-wideband signal in the time domain consists of pulses with the periodicity T. In general, a periodicity with T in the time domain leads to a frequency spectrum with discrete lines which are separated by the distance $\triangle f = 1/T$. An unmodulated UWB signal shows discrete energy peaks in the frequency domain that can violate the given spectral mask. Modulation with data reduces the strong periodicity of the signal in the time domain, but in most cases this will not be sufficient to achieve a flat spectrum in the relevant frequency range (e.g. between 3.1 and 10.6 GHz, assuming the FCC mask). More flatness of the spectrum is achieved by introducing a pseudo-random timing offset to the pulse w.r.t. the nominal position. The generation of pseudo-random offsets can be realized by a time-hopping code. In this case, T is divided into N_{TH} time slots, and a code defines in which time slot the modulated pulse has to be set. The length of the time-hopping (TH) time slot T_{TH} is

$$T_{\text{TH}} = \frac{T}{N_{\text{TH}}}. \tag{2.36}$$

In general, N_{TH} is a power of 2 (e.g. 2, 4, 8, 16, 32, ...). Increasing N_{TH} leads to more flatness in the frequency domain.

When PPM is used, the TH coding is constrained by the condition

$$T_{\text{TH}} < T_{\text{PPM}} + T_{\text{p}}. \tag{2.37}$$

Random offsets can also be achieved by a two-step approach. First, the pulse repetition time is subdivided into (for example) a small number of time slots, and a certain time slot

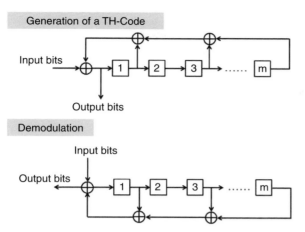

Figure 2.21 Generation of a TH-Code by a shift register at the Tx side and demodulation at the Rx side. ©2010 KIT Scientific Publishing; reprinted with permission from [168].

is selected as just described (coarse TH code). Then, a second code (fine TH) introduces very small offsets around this position. The pulses tremble around the nominal position like a jitter, which leads to a high degree of flatness in the frequency domain. A detailed analysis of the power spectral density as a function of the modulation and the TH coding can be found in [121, 180].

2.7.1 Generation of time-hopping codes

The code used for the TH coding should be easy to implement and should lead to a flat spectrum. Gold codes and maximum length codes (also called m-sequences) are suitable candidates [111]. The following consideration assumes m-sequences. A TH code is achieved by an m-bit shift register that is based on a primitive polynomal, see Fig. 2.21. The m-bit shift register generates $2^m - 1$ different codes, excluding the zero vector. A series of m zeros in the shift register is excluded since this state cannot generate other states. A code 010 resulting from a 3-bit shift register can be used to shift the position of a modulated pulse by $0 \cdot 2^0 + 1 \cdot 2^1 + 0 \cdot 2^2 = 2$ time slots. The total number of TH slots per pulse repetition time, called N_{TH}, is defined by

$$N_{TH} = 2^m. \tag{2.38}$$

Figure 2.22 shows a PPM modulated signal including TH coding in the time domain for a bit value of 1. TH coding can also be used to separate different users by a code.

2.8 Basic transmitter architectures

A transmitter model for impulse radio transmission is shown in Fig. 2.23. The pulse repetition time is generated by a clock oscillator. In the time domain, PPM or TH

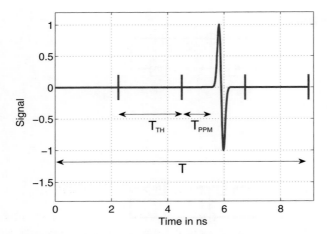

Figure 2.22 UWB pulse including TH coding; assumptions: bit value 1 and a TH code of 010. ©2010 KIT Scientific Publishing; reprinted with permission from [168].

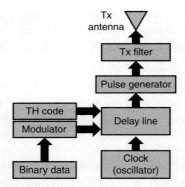

Figure 2.23 Transmitter model (Impulse Radio transmission). ©2010 KIT Scientific Publishing; reprinted with permission from [168].

coding require the pulses to be set at offsets w.r.t. to the clock. This can be realized by a programmable delay line. Finally, a pulse shape generator generates ultra-wideband pulses in the time domain that are radiated by the Tx antenna at defined timestamps.

2.9   Basic receiver architectures

A UWB signal can be demodulated by coherent and incoherent methods. A good overview can be found in [139, 141, 161]. Here, a brief overview is presented. In general, a suited demodulation scheme depends on the application, the hardware characteristics, and the trade-offs between cost, complexity and performance.

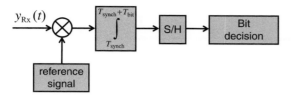

Figure 2.24 Coherent receiver. ©2010 KIT Scientific Publishing; reprinted with permission from [168].

2.9.1   Coherent receiver for on–off keying and pulse position modulation

Coherent demodulation calls for a good synchronization, which increases the complexity and demands for high-stability oscillators. This leads to an improved performance compared to an incoherent demodulation in AWGN channels [141]. The principle of a coherent receiver for OOK or PPM is shown in Fig. 2.24. Basically, the received signal $y_{Rx}(t)$ is multiplied with a known reference signal, also known as the template signal. The product is integrated over time, where the total integration time corresponds to the bit duration. After a sample and hold (S/H) circuit, a threshold detector performs the bit decision.

In the following, PPM in particular is considered in more detail. The received signal is multiplied by a synchronized reference signal (also known as the template signal) and integrated over the bit duration. If a bit is represented by N pulses, the bit duration is $N \cdot T$. The result of the integration is compared to a threshold of zero for the bit decision. After a reset to zero, the procedure is repeated until all bits are demodulated. For a better understanding, some comments on the reference signal are given. The pulse shape of the reference signal (also known as the template pulse) is the difference signal between the signal for a binary 0 and 1. Ideally, the difference signal should be based on the distorted received signal to maximize the SNR. However, this would require the knowledge of the system transfer function and hence increase the complexity. Therefore, it is a common approach to use the undistorted pulse shape for the template pulse. The template pulse must be placed at multiples of the pulse repetition time T. In the case of an additional TH coding at the Tx side, the code must also be known at the Rx side. The template pulses are then further delayed by values defined by the TH code. Figure 2.25 visualizes a possible received PPM modulated signal using Gaussian monocycle pulse shapes with $T_{PPM} > T_p$, where T_p is the duration of the Gaussian monocycle. In the case of PPM, the template signal consists of the original pulse itself and a (by T_p) delayed and inverted version of the original pulse. This can be seen in the middle plot (of the three) in Fig. 2.25. The lower plot results from multiplication, integration versus time, and a reset to zero after the bit duration. It can be clearly observed that the integrated values before the reset are either positive or negative, which determines the bit decision. Even if the received signal disappears in the noise, the ideal threshold for the bit decision remains zero. In this case, the values before the reset differ in general only slightly from zero, which leads to a worse system performance.

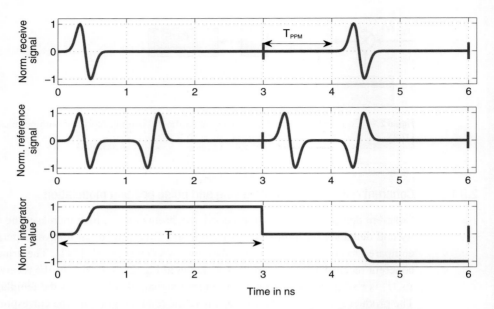

Figure 2.25 Principle of coherent demodulation in the case of PPM. ©2010 KIT Scientific Publishing; reprinted with permission from [168].

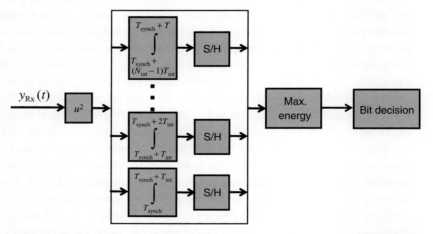

Figure 2.26 Incoherent receiver for PPM: detection of the maximum integrator value (MAX method). ©2010 KIT Scientific Publishing; reprinted with permission from [168].

2.9.2   Incoherent receiver for pulse position modulation

The possibility to detect a PPM modulated signal by an incoherent receiver is shown in Fig. 2.26. The time slot with the maximum energy is chosen for the detection [129, 140, 162]. In contrast to coherent demodulation, there is no template signal. Instead, the received signal is squared and integrated. Since the binary information is stored in the temporal position, the idea is to divide the pulse repetition time into a set of time slots

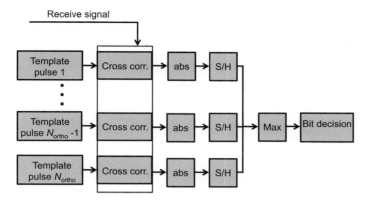

Figure 2.27 Receiver for modulation with N_{ortho} orthogonal pulses. ©2010 KIT Scientific Publishing; reprinted with permission from [168].

and to find the slot that contains the maximum signal energy. For example, without a PPM offset, the maximum energy is found inside the first slot, while the presence of PPM shifts the maximum energy to another time slot.

2.9.3   Receiver for orthogonal modulation

A receiver for orthogonal modulation is shown in Fig. 2.27. Here, the received signal is correlated with N_{ortho} (e.g. eight) different template pulses, where each template pulse represents a different bit sequence with $\log_2 N_{\text{ortho}}$ bits (e.g. 2 bits). The template pulse, which then leads to the maximum correlation coefficient, is used for the bit decision.

3 UWB antennas

Grzegorz Adamiuk, Xuyang Li and Werner Wiesbeck

3.1 UWB antenna measurement methods

The methods for measuring antenna transfer functions with a network analyzer presented in the following use typically available measurement equipment for far-field antenna measurements with enhanced calibration procedures. The latter are necessary in order to obtain stable and consistent phase information for the transfer function in co- and cross-polarization. These methods complement the standardized antenna measurements known from [68]. A validation of the methods is presented with the prediction of a time domain impulse transmission, which is measured independently with a fast-pulse generator and an oscilloscope.

A prerequisite for the measurements is a stable measurement setup inside an anechoic chamber, which provides low reflections from the walls within the frequency range of operation. If multiple reflections cannot be avoided, the possibility of time gating for valid signals of the additional delay – due to the path lengths of wall or floor reflected signals – needs careful consideration. If the expected τ_r or even τ_{FWHM} overlap with the multiple reflections, the impulse response $\mathbf{h}_{AUT}$ of the antenna under test (AUT) can no longer be measured accurately. Furthermore, far-field conditions are necessary in order to obtain a distance-independent impulse response estimation $\mathbf{h}_{AUT}$. The frequency domain far-field criterion $r > (2D^2)/\lambda$ (according to [24]) then applies for the highest frequency, i.e. the smallest wavelength. Near-field antenna ranges for impulse response measurements are still an open topic, however this seems to be a straightforward application of the corresponding near-to-far-field theorems. Additionally the measurement frequency or time sampling rate needs to be safely within the Shannon theorem boundaries in order to avoid spoiling the measurement results with aliasing. Last but not least the dynamic range of the measurement setup needs to be large enough for the expected signal variation, and the test fixtures need to be stable enough for reproducible measurements with respect to angular resolution.

A typical configuration for the measurement setup is shown in Fig. 3.1 with a distance $r_{TxRx} = 2.64$ m between the reference antenna and the antenna under test. It is important for the measurement that the axis of the antenna rotation coincides with the center of radiation of the antenna. Otherwise, the impulse response will show an additional delay offset, which depends on the turning angle.

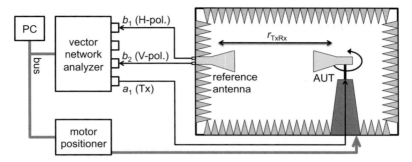

Figure 3.1 Typical configuration for a standard antenna test range with dual-polarization capability.

3.1.1 Calibration and substitution method

In general two different calibration methods are known for antenna measurements. First, the absolute methods where the transmission is calibrated with unknown antennas (for instance, two identical but unknown antennas, or three non-identical unknown antennas). Second, referencing the unknown antenna to a known antenna standard. The latter is also known as the substitution method (see also [67] with respect to measurement methods for antenna gain measurements). Every substitution method needs to use at least once an absolute measurement method for obtaining the calibration data of the antenna standard (the so called "golden device").

The general measurement setup is shown in Fig. 3.1. With this setup only transmission is measured. The input reflection coefficient S_{11} of the antenna is measured separately with a calibrated network analyzer. The transmission coefficient is referenced to the systems transmission characteristic $S_{21,\text{sys}}$ which is obtained when connecting the terminals of transmit and receive antennas. The transmission factor S_{21} is calculated according to

$$S_{21} = \frac{S_{21,\text{raw}}}{S_{21,\text{sys}}} . \tag{3.1}$$

The system transmission characteristic $S_{21,\text{sys}}$ is typically flat in magnitude and linear in phase in order to obtain good measurement results. Multiple reflections between receiving antenna and receiver are usually no problem since the receiver is very well matched to the 50 Ω transmission line.

The substitution method uses a known antenna standard for which the antenna transfer function $H_{\text{gold,co}}$ in co-polarization is known within the frequency range of interest, and which is different from the reference antenna used for receiving. First the raw transmission coefficient $S_{21,\text{gold,raw}}$ is measured with this substitution standard by placing the "golden device" on the positioner instead of the AUT. Then the AUT is measured with the same setup and the transfer function of the AUT is calculated according to (3.2):

$$H_{\text{AUT,co}} = \frac{S_{21,\text{AUT,raw}}}{S_{21,\text{gold,raw}}} H_{\text{gold,co}} \tag{3.2}$$

To select the substitution, standard antennas with a flat transfer function within the frequency range of interest are preferred. Notches or increased ringing will particularly distort the measurement.

3.1.2 Two-antenna method

In the following, the basic two-antenna method for the measurement of the antenna impulse response is described. It uses two antennas of the same type and fabrication that can be assumed to be identical. Then the transmission is set up with the antennas oriented in co-polarization and the main beam direction is towards each other. For the measurement of the cross-polar components and the compensation of cross-polarization properties, a second measurement with one antenna turned 90° is necessary. The polarimetric measurement has higher requirements regarding the assumption that the antennas are identical, since the cross-polar components are often weak and can be influenced by the test fixture.

Based on (2.7) and (2.10), the transmission from one antenna to the other can be modeled as [158]

$$S_{21}(f) = \frac{U_{Rx}(f)}{U_{Tx}(f)} = \sqrt{\frac{Z_{C,Rx}}{Z_{C,Tx}}} \cdot \frac{e^{-j\omega r_{TxRx}/c_0}}{2\pi r_{TxRx}c_0} \cdot j\omega \mathbf{H}_{Rx}^T(f, \theta_{Rx}, \psi_{Rx}) \cdot \mathbf{H}_{Tx}(f, \theta_{Tx}, \psi_{Tx})$$

$$(3.3)$$

$$s_{21}(t) = \sqrt{\frac{Z_{C,Rx}}{Z_{C,Tx}}} \cdot \frac{\delta(t - \frac{r_{TxRx}}{c_0})}{2\pi r_{TxRx}c_0} * \frac{\partial}{\partial t} \mathbf{h}_{Rx}^T(t, \theta_{Rx}, \psi_{Rx}) * \mathbf{h}_{Tx}(t, \theta_{Tx}, \psi_{Tx}). \quad (3.4)$$

Note that the antenna transfer functions **H** and their transient response functions **h** are each presented in their individual spherical coordinate systems.

Two-antenna method without cross-polarization compensation
With the basic measurement of two antennas pointing in co-polarization towards each other, and $Z_{C,Tx} = Z_{C,Rx}$, the expression in (3.3) reduces [151] to

$$S_{21} = \frac{e^{-j\omega r/c_0}}{2\pi r_{TxRx}c_0} j\omega H_{ref,co}^2(\theta_{mb}, \psi_{mb}), \quad (3.5)$$

where θ_{mb} and ψ_{mb} denote the main beam direction of the two identical antennas. It is emphasized that mismatch at the antenna ports is automatically taken into account, since the network analyzer measurement fixes the characteristic impedance Z_C to the characteristic impedance of the measurement system, which is typically equal to 50 Ω. With this expression, the transfer function of the two identical antennas is calculated as follows:

$$H_{ref,co} = \sqrt{\frac{2\pi r_{TxRx}c_0}{j\omega} S_{21} \exp(j\omega r_{TxRx}/c_0)}. \quad (3.6)$$

Note that for the numerical evaluation of the square root with $\sqrt{z} = \sqrt{|z|} \exp(j\phi(\omega)/2)$ it is necessary to use an unwrapped (i.e. continuous) phase over frequency $\phi(\omega)$.

The transformation into the time domain is achieved with an inverse discrete Fourier transform [98]

$$h_{\text{ref,co}}^{+}(k\Delta t) = \frac{1}{N\Delta t} \sum_{n=0}^{N-1} H_{\text{F}}(k\Delta f) H_{\text{ref,co}}^{+}(n\Delta f) \cdot \exp\left(\frac{\text{j}2\pi k n}{N}\right) \tag{3.7}$$

$$h_{\text{ref}}(k\Delta t) = \Re\left\{h^{+}(k\Delta t)\right\}. \tag{3.8}$$

The practical measurements are obtained for positive frequencies. The analytic impulse response $h_{\text{ref,co}}^{+}(k\Delta t)$ in (3.7) is complex. Using the relations of the Hilbert transform in (3.8) the real part of the analytic signal yields the antenna impulse response $h_{\text{ref}}(k\Delta t)$. Note that the spectrum of the analytic signal $H_{\text{ref,co}}^{+}(n\Delta f)$ has only positive values, which have double the amplitude. The frequency range of the measurement is chosen according the band of the available reference antenna and is kept the same for all measurements.

Two-antenna method with cross-polarization compensation
For the purpose of cross-polarization compensation, two measurements are conducted. First, the two identical antennas are oriented in co-polarization with the result $S_{21,\text{co}}$. Second, one of the two antennas is rotated $90°$ around the main beam axis and the transmission $S_{21,\text{x}}$ for the cross-polarized case is obtained. Applying (3.3) yields

$$S_{21,\text{co}} = \gamma\left(H_{\text{co}}^{2} - H_{\text{x}}^{2}\right) \tag{3.9}$$

$$S_{21,\text{x}} = -2\gamma\, H_{\text{co}} H_{\text{x}} \tag{3.10}$$

with

$$\gamma = \sqrt{\frac{Z_{\text{C,Rx}}}{Z_{\text{C,Tx}}}}\, \frac{\exp(-\text{j}\omega r_{\text{TxRx}}/c_0)}{2\pi r_{\text{TxRx}} c_0}\, \text{j}\omega. \tag{3.11}$$

This equation system is solved for H_{co} and H_{x}:

$$H_{\text{co}} = \sqrt{\frac{S_{21,\text{co}}}{\gamma}\left(1 + \sqrt{1 + \frac{S_{21,\text{x}}^{2}}{S_{21,\text{co}}^{2}}}\right)} \tag{3.12}$$

$$H_{\text{x}} = -\frac{S_{21,\text{x}}}{2\gamma\, H_{\text{co}}}. \tag{3.13}$$

3.1.3   Three-antenna method

The three-antenna method enables a direct measurement of the antenna transfer function without assuming two antennas being identical. However all antennas in this procedure need to cover the frequency range of interest with sufficient performance. This means that their transfer functions should not have notches. After their transfer function is determined, they can be used as reference antennas or golden devices according to Section 3.1.1.

Three-antenna method without cross-polarization compensation

For the three-antenna method without cross-polarization compensation, the antennas' transfer functions in co-polarization are denominated with $H_{co,1}$, $H_{co,2}$ and $H_{co,3}$. The calibrated transfer functions from antenna i to antenna j are given by S_{ji}. To calculate the antenna transfer function, the transmissions S_{21}, S_{31} and S_{23} are measured. Using the formulation for γ from (3.11), these transmissions are modeled as follows:

$$\frac{S_{21}}{\gamma_{21}} = H_{co,2} \, H_{co,1}$$

$$\frac{S_{31}}{\gamma_{31}} = H_{co,3} \, H_{co,1} \qquad (3.14)$$

$$\frac{S_{23}}{\gamma_{23}} = H_{co,2} \, H_{co,3}$$

with

$$\gamma_{ij} = \sqrt{\frac{Z_{C,Rx}}{Z_{C,Tx}}} \, \frac{\exp(-j\omega r_{TxRx,ij}/c_0)}{2\pi r_{TxRx,ij} c_0} \, j\omega.$$

This set of equations is solved for $H_{co,i}$:

$$H_{co,1} = \frac{S_{21} S_{31} \gamma_{23}}{S_{23} \gamma_{21} \gamma_{31}}$$

$$H_{co,2} = \frac{S_{21} S_{23} \gamma_{31}}{S_{31} \gamma_{21} \gamma_{23}} \qquad (3.15)$$

$$H_{co,3} = \frac{S_{23} S_{13} \gamma_{12}}{S_{12} \gamma_{23} \gamma_{13}}.$$

Three-antenna method with cross-polarization compensation

The cross-polarization compensation of the two-antenna method presented in Section 3.1.2 may be conveyed to a three-antenna method. Therefore every measurement of a pair of antennas is conducted in co- and cross-polarization orientation. This pair of measurements is modeled according to:

$$\begin{pmatrix} S_{ij,co} \\ S_{ij,x} \end{pmatrix} = \gamma_{ij} \begin{pmatrix} H_{co,i} & -H_{x,i} \\ -H_{x,i} & -H_{co,i} \end{pmatrix} \begin{pmatrix} H_{co,j} \\ H_{x,j} \end{pmatrix}. \qquad (3.16)$$

If $\mathbf{H}_i = (H_{co,i} \quad H_{x,i})^{\mathrm{T}}$ is known, then the equation is resolved to:

$$\begin{pmatrix} H_{co,j} \\ H_{x,j} \end{pmatrix} = \frac{1}{\gamma_{ij} \left(H_{co,i}^2 + H_{x,i}^2 \right)} \begin{pmatrix} H_{co,i} & -H_{x,i} \\ -H_{x,i} & -H_{co,i} \end{pmatrix} \begin{pmatrix} S_{ij,co} \\ S_{ij,x} \end{pmatrix}. \qquad (3.17)$$

From the three pairs of antennas, we get six measurements for the six unknown transfer function components. In [74] the complex polarization ratios are ξ_i

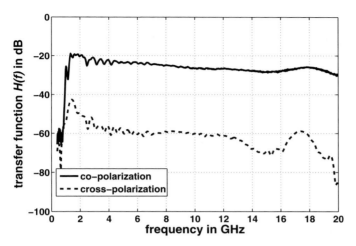

Figure 3.2 Transfer function of an example reference broadband horn antenna (EM Systems, model 6100) in co- and cross-polarization. ©2007 IHE; reprinted with permission from [158].

and M_{ij}:

$$\xi_i = \frac{H_{\text{co},i}}{H_{\text{x},i}} \tag{3.18}$$

$$M_{ij} = \frac{S_{ij,\text{x}}}{S_{ij,\text{co}}}. \tag{3.19}$$

The polarization ratio ξ_3 is calculated from M_{12}, M_{13} and M_{23} according to [74]:

$$\xi_3 = \frac{\sqrt{M_{12}^2 + 1}\sqrt{M_{13}^2 + 1}\sqrt{M_{23}^2 + 1} + (M_{13} - M_{12}) M_{23} - M_{12} M_{13} - 1}{(M_{12} M_{13} + 1) M_{23} + M_{13} - M_{12}}. \tag{3.20}$$

The polarization ratios ξ_i do not give an indication of the absolute values of the components H_{co} and H_{x}. This is achieved by rewriting (3.17) for the transmission with ξ_i, which is then solved for $H_{\text{x},3}$ [158]. The algebraic solution takes advantage of the freeware package Maxima [100]:

$$H_{\text{x},3} = \frac{\sqrt{S_{23,\text{x}}^2 + S_{23,\text{co}}^2}}{\sqrt{\xi_3^2 + 1}} \left(\frac{\xi_3 S_{31,\text{co}}}{\xi_3 S_{21,\text{x}} S_{23,\text{x}} - S_{21,\text{co}} S_{23,\text{x}} + \xi_3 S_{21,\text{co}} S_{23,\text{co}}} \right.$$

$$\left. - \frac{S_{31,\text{x}}}{\xi_3 S_{21,\text{x}} S_{23,\text{x}} - S_{21,\text{co}} S_{23,\text{x}} + S_{21,\text{x}} S_{23,\text{co}} + \xi_3 S_{21,\text{co}} S_{23,\text{co}}} \right)^{1/2} \tag{3.21}$$

$$H_{\text{co},3} = \xi_3 H_{\text{x},3}. \tag{3.22}$$

Again, the complex root is evaluated along a branch with steady increasing phase over frequency. With $H_{\text{x},3}$, all six components of $\mathbf{H}_1$, $\mathbf{H}_2$ and $\mathbf{H}_3$ are computed from (3.17). The resulting transfer functions of an example of broadband ridged horn antenna (EM Systems, model 6100) for co- and cross-polarization are shown in Fig. 3.2.

3.1.4   Direct measurement with one reference standard

A very practical measurement of the antenna transfer function is the direct measurement of the transmission from the AUT to the known reference antenna at a known distance r_{TxRx}. Neglecting the cross-polarization properties, the transmission is modeled according to:

$$S_{21}(\theta, \psi) = \frac{e^{-j\omega r_{\text{TxRx}}/c_0}}{2\pi r_{\text{TxRx}} c_0} \, j\omega H_{\text{AUT,co}}(\theta, \psi) H_{\text{ref,co}}(\theta_{\text{Tx}}, \psi_{\text{Tx}}). \qquad (3.23)$$

This yields $H_{\text{AUT,co}}$ as

$$H_{\text{AUT,co}}(\theta, \psi) = 2\pi r_{\text{TxRx}} c_0 e^{+j\omega r_{\text{TxRx}}/c_0} \frac{S_{21}(\theta, \psi)(j\omega H_{\text{ref,co}})^*}{\left| j\omega H_{\text{ref,co}} \right|^2 + K}. \qquad (3.24)$$

In this expression the complex division is obtained with a simplified Wiener filter [119], where the constant K is of the order of magnitude of the noise power. The simplified Wiener filter suppresses an amplification of the noise level $j\omega H_{\text{ref,co}}^2$ at frequencies where the magnitude of $H_{\text{ref,co}}$ is low. Further processing follows the same scheme as shown in (3.7). With measurements for co- and cross-polarization, this direct method is also usable for the polarimetric determination of the AUT's transfer function according to (3.17).

3.1.5   Verification in the time domain

Using (3.3) for the UWB free-space transmission, the transfer function H_{AUT} of the antenna under test is extracted in the frequency domain from the network analyzer measurement of the transmission S_{21}. In the following this result is compared to the direct time domain measurement, with a fast pulse as source and a digital sampling oscilloscope as receiver, based on (3.4). For the source, a PSPL 3600 (Picosecond Pulse Labs) has been used. The receiver has been the Agilent Infiniium DCA with 40 GSa/s and 12 bit resolution. The receiver has been triggered by the received signal with an average of 16–64 single shots. The exciting pulse has a width of $\tau_{\text{FWHM}} = 78$ ps. The signal in Fig. 3.3 is measured, including the connection losses of a 1 m coaxial cable. Both measurements have been performed in an anechoic chamber.

In Fig. 3.4, the comparison of the modeled received voltage (according to (3.5)) and the directly measured received voltage is shown. The modeled data is obtained by a convolution of the measured frequency domain transfer function H_{Tx} of the applied antenna with the exciting signal in the time domain (cf. Fig. 3.3). The perfect agreement of model and measurement in Fig. 3.3 proves the correctness of the previous formulas.

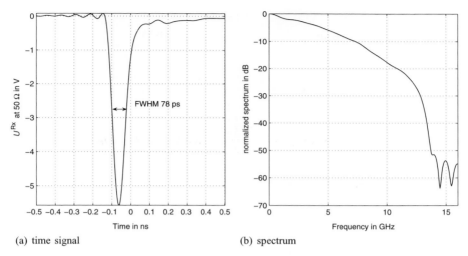

(a) time signal   (b) spectrum

Figure 3.3 (a) Source pulse and (b) corresponding spectrum of the pulse generator (Picosecond Pulse Labs model PSPL 3600). ©2007 IHE; reprinted with permission from [158].

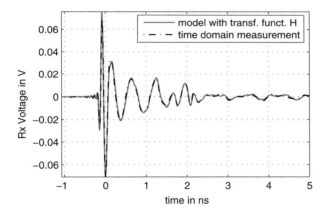

Figure 3.4 Comparison of the model according to (3.5) and a direct time domain measurement with the pulse from Fig. 3.3. The figure shows the transmission between two ridged broadband horn antennas. ©2007 IHE; reprinted with permission from [158].

3.2   UWB radiator design

3.2.1   UWB antenna principles

The radiation of guided waves has been discussed intensively in the past. It is the common understanding that the key mechanism for radiation is charge acceleration [105, 106]. The question to answer for UWB is: what kind of structures facilitate the charge acceleration over a very wide bandwidth? The ultra-wide bandwidth radiation is based on a few principles:

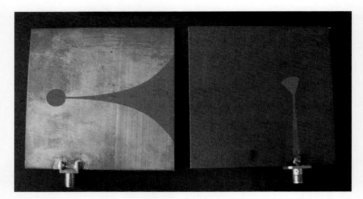

Figure 3.5 Aperture coupled Vivaldi antenna. Left: top view; right: bottom view with feed line; substrate size 75×78 mm^2. ©2009 IEEE; reprinted with permission from [179].

- traveling wave structures
- frequency-independent antennas (angular constant structures)
- self-complementary antennas
- multiple resonance antennas
- electrically small antennas.

In most cases, the radiation starts where the electric field connects $180°$ out of phase currents with half a wavelength spacing. This is not the case for electrically small antennas for which the impedance then varies strongly with frequency. Many antennas radiate by a combination of two or more of the above principles and cannot therefore be simply classified. In the following the relationships between the radiation principles and the properties of the antennas are discussed. Each explanation of the radiation phenomenon is supported by an example of an antenna.

3.2.2   Traveling wave antennas

Traveling wave antennas offer a smooth, almost unrecognizable transition for the guided wave, with the fields accelerated to free-space propagation speed c_0. Typical antennas are tapered wave guide antennas [150], for example the horn antenna (see Fig. 2.14) or the Vivaldi antenna (see Fig. 3.5). Other examples of radiating traveling wave structures are the slotted waveguide and the dielectric rod antenna. Here we will use the Vivaldi antenna as an example, for which different feed structures like microstrip line, slotline, and antipodal can be applied.

    The Vivaldi antenna guides the wave from the feed in a slotline to an exponential taper. The exponential taper is a priori wideband. A typical structure, etched on a dielectric substrate, is shown in Fig. 3.5. The Vivaldi is fed at the narrow side of the slot. For UWB the major tasks are the wideband, frequency-independent feed and the slotline terminations. The feed shown here is designed with a Marchand balun network with aperture coupling. Non-resonant aperture coupling is usually a good choice for UWB feed structures as it allows for impedance matching over a wide range of frequencies. A

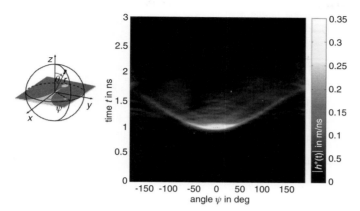

Figure 3.6 Measured impulse response $h^+(t, \psi)$ (co-pol) of the Vivaldi antenna of Fig. 3.5 in the E-plane, $\theta = 90°$. ©2009 IEEE; reprinted with permission from [179].

stub, and the slotline by a circular shaped cavity, terminate the microstrip feed line. The antenna can be designed relatively compact. The propagation velocity v on the structure changes from the slotline wave velocity v_{sl} to c_0 at the end of the taper. It varies only slightly with frequency.

The Vivaldi antenna's time domain transient response in the E-plane is shown in Fig. 3.6. This representation is unusual for narrowband antennas. It displays the impulse distortion by the antenna in time t versus the E-plane angle ψ. The Vivaldi antenna has a rather low distortion compared to other UWB antennas. The high peak value ($p = 0.35$ m/ns) and the short duration of the transient response envelope ($\tau_{FWHM} = 135$ ps), stand for very low dispersion and ringing (see Fig. 2.14). The ringing of the antenna is due to multiple reflections at the substrate edges and parasitic currents along the outer substrate edges. The ringing can be reduced by enlarging the transverse dimensions of the antenna by metal flares or chokes. Absorbing material around the substrate edges reduces the ringing without influencing the other characteristics of the transient response. The slightly asymmetric impulse response results from the feed line. The frequency and angle-dependent gain $G(f, \theta = 90°, \psi)$ in Fig. 3.7 is calculated from the measured directional transfer function $H(f, \theta = 90°, \psi)$. The gain is quite constant versus frequency in the main beam direction. The maximum gain G_{max} is 7.9 dBi at 5.0 GHz; at the lower frequencies (close to 3 GHz) small resonances are visible. The average gain $\bar{G}$ in the FCC frequency band is 5.7 dBi. The main parameters of the Vivaldi antenna are summarized in Table 3.1. The Vivaldi antenna is well suited for direct planar integration and also for UWB antenna arrays for radar and communications. In the past it has also been used for special cases of high power radiation.

3.2.3   Frequency-independent antennas

Rumsey investigated the fundamentals of frequency-independent antennas in the 1960s [146]. He observed that a scaled version of a radiating structure must exhibit the same

Table 3.1 UWB parameters of the Vivaldi antenna of Fig. 3.5 in the main beam direction.

Parameter	Value
p_{max} (m/ns)	0.35
τ_{FWHM} (ps)	135
$\bar{G}$ (dBi)	5.7
G_{max} (dBi)	7.8
$\tau_{r=0.22}$ (ps)	150

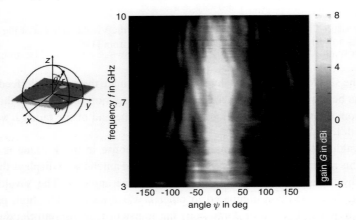

Figure 3.7 Measured gain $G(f, \theta = 90°, \psi)$ of the Vivaldi antenna of Fig. 3.5 in the E-plane versus frequency. ©2009 IEEE; reprinted with permission from [179].

characteristics as the original one, when fed with a signal whose wavelength is scaled by the same factor. As a consequence, if the shape of an antenna is invariant to physical scaling, its radiating behavior is expected to be independent of frequency. The typical realization is an angular constant structure, described only by angles. It must be noted that the independence from frequency does not necessarily refer to the input impedance of the structure. In order to obtain a constant input impedance, additional principles such as those described in Section 3.3 have to be applied.

The scaling usually involves constant angles. It is possible to define a "truncation principle" to apply this concept to the practical case, where the size of any physical object is obviously finite [101]. In fact, provided that the overall current on the antenna tends to decrease (due to the radiation) when moving away from the feeding point, it is possible to define a limited "active" region, where the current falls below relevant values. If the actual, finite antenna contains this region, it can be assumed that the truncation of the geometry does not modify the behavior of the antenna around the chosen wavelength. A typical example of a frequency-independent antenna is the biconical antenna [24].

Figure 3.8 Left: aperture coupled bow-tie antenna: left-hand side – bottom view with feed line; right-hand side – top view (rule is showing centimeters); right: symmetric fed bow-tie antenna with balun. ©2009 IEEE; reprinted with permission from [179].

A planar example of the biconical antenna is the bow-tie antenna. The antenna structure consists of two triangular metal sheets (see Fig. 3.8). They are usually fed by a symmetric (twin) line, which is matched to the feed point impedance. In the case of an asymmetric feed line (like coaxial or microstrip lines) a balun transformer is needed. The bow-tie antenna has, for the FCC UWB frequency band, reasonable dimensions. The application of aperture feed and further optimizations allow a very compact design.

The aperture coupled bow-tie antenna consists of two triangular radiating patches, one of which serves as a ground plane for the tapered microstrip feed line that ends with a broadband stub (see Fig. 3.8). The feeding structure couples the energy from an asymmetric microstrip line to the radiating bow-tie elements through the aperture formed by the tips of the triangles. Therefore the antenna is called an aperture coupled bow-tie (ACB) antenna. This feeding technique is basically similar to the operation of the well-known microstrip–slotline transitions with a Marchand balun. Almost no additional ringing is introduced by this coupling mechanism. The pulses on the radiating elements are traveling faster than those on the line, due to the lower effective $\varepsilon_{r,eff}$. This is compensated for by the fact that the stub length is shorter than the length of the radiating elements.

The ACB antenna has a nearly omnidirectional radiation pattern in the H-plane (Fig. 3.9). Therefore this type of antenna (which can be manufactured quite small) can be used, for example, in communications for mobile devices. Figure 3.10 shows the measured impulse response $|h^+(t)|$ of the antenna in Fig. 3.8. The almost omnidirectional radiation in the H-plane is well visible, accompanied by a small ringing of the antenna.

Other types of antennas with frequency-independent characteristics might be some versions of logarithmic-periodic antennas or spiral antennas [58]. Although these antennas can show frequency-independent characteristics, they are based on a different design principle and their properties are different when compared to those described above. In general, antennas may combine more than one radiation principle, and they may change the radiation principle versus frequency.

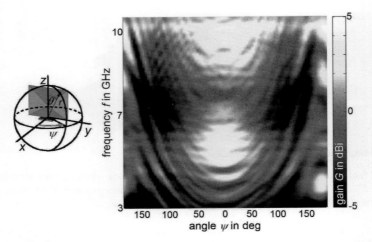

Figure 3.9 Measured gain $G\,(f, \theta = 90°, \psi)$ of the bow-tie antenna in the H-plane versus frequency.

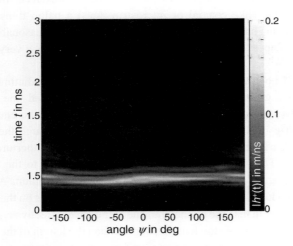

Figure 3.10 Measured impulse response $h^+\,(t, \psi)$ of the bow-tie antenna in the H-plane. ©2009 IEEE; reprinted with permission from [179].

3.2.4 Self-complementary antennas

Self-complementary antennas are characterized by a self-complementary metalization [113]. This means that metal can be replaced by dielectric and vice versa without changing the antenna's structure (Fig. 3.11). The behavior of self-complementary structures can be analyzed by applying Babinet's principle [27]. This results in an invariant input impedance of

$$Z_{\text{in}} = Z_{\text{F0}}/2 = 60\pi \ \Omega \tag{3.25}$$

with Z_{F0} being the free-space impedance. Self-complementary structures only guarantee a constant input impedance, but not necessarily constant radiation characteristics,

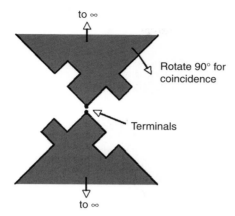

Figure 3.11 Truncated fractal antenna to show the principle of self-complementary antennas. ©2009 IEEE; reprinted with permission from [179].

Figure 3.12 Self-complementary antennas. Left: 2-arm logarithmic spiral antenna; right: sinuous antenna. ©2009 IEEE; reprinted with permission from [179].

independent of frequency. It is also possible to design structures that are similar to self-complementary structures, but have an unbalanced ratio of metalized to non-metalized areas. These structures exhibit an input impedance which is nearly constant versus frequency, but different from $Z_{F0}/2$ [28]. For an exact description of self-complementary antennas, the reader is referred to [39]. Typical candidates are the 90° bow-tie antenna, the sinuous antenna, the logarithmic spiral antenna [42] or some fractal antennas [178].

An example of a 2-arm logarithmic spiral antenna is shown in Fig. 3.12. This antenna realizes the principle of frequency independence; the metalization is only defined by angles and it follows Babinet's principle. The two arms of the antenna are fed in the center with a symmetric line with an impedance of $Z_L = 60\pi \ \Omega$. When properly designed, the logarithmic spiral antenna radiates where the two arms are spaced by $\lambda/2$, i.e. where the circumference is $\lambda\pi/2$. With a good structural design, the antenna can be made broadband up to a bandwidth of several 100%.

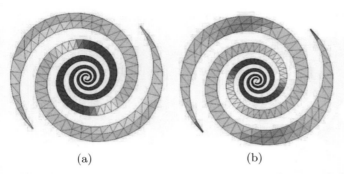

The principle of radiation can be seen by plotting the current distribution into the antenna of Fig. 3.13. The diameter of the outer arms reaches 40 cm in this case. In Fig. 3.13(a) the antenna is excited with the frequency of 300 MHz and in Fig. 3.13(b) with 450 MHz. It can be noted that at lower frequencies, where the wavelength is larger, the high current amplitudes occur at a larger diameter on the spiral than in the case of the higher frequency of 450 MHz. The "vanishing" of the currents outside the maximum currents indicates that energy has been radiated.

The logarithmic spiral antenna is a directional antenna with two main beams orthogonal to the spiral plane. In general, the radiated wave is circularly polarized. The polarization in the two opposite radiation directions is orthogonal (i.e. left- and right-handed circularly polarized). In practical applications, radiation is desired in only one direction, and for that purpose one of the beams is usually suppressed by absorbing material. The position of the radiating area is frequency dependent, as is the time delay. This results in a broadening and a smaller peak value of the antenna impulse response, compared to (for example) the Vivaldi antenna.

As another example of a frequency-independent antenna, the Archimedean spiral antenna is analyzed. This antenna has, compared to the logarithmic spiral antenna, constant linewidth and spacing, and these are usually identical. Due to the close proximity of adjacent lines they couple strongly. This causes radiation where adjacent lines are in phase, i.e. where the circumference is λ. To analyze the transmitted pulse characteristics, a linearly polarized electric field probe in the simulation tool is set in the far field of the Archimedean spiral antenna. The electric field probes are arranged along the x-, y- and z-axes, whereas the antenna is positioned in the x–y-plane. The received electric field $e(t)$ of a simulation of this configuration excited with a $\tau_{\text{FWHM}} = 88$ ps Gauss pulse is shown in Fig. 3.14. The radiated UWB signal has a strong, short peak with a reasonable ringing. Since the radiated pulse is circularly polarized, both $e_x(t)$ and $e_y(t)$ components are present. The $e_z(t)$ components are not excited since the radiated wave is a TEM wave.

To ensure circular polarization of a spiral antenna for UWB systems with very short pulses, its pulse duration must be long enough to cover 360° when traveling around the spiral arms.

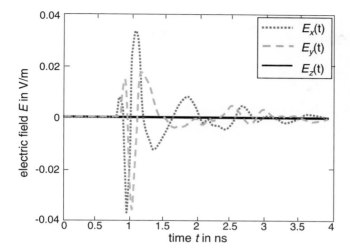

Figure 3.14 Simulation of the received electric field $e_{Rx}(t)$ from an Archimedean spiral antenna (x–y plane) with linear polarized electric field probes in x-, y- and z-directions; input pulse is a Gaussian pulse with $\tau_{FWHM} = 88\,\text{ps}$. ©2009 IEEE; reprinted with permission from [179].

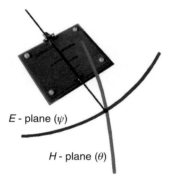

Figure 3.15 Log-per antenna with a coaxial connector feeding the inner triplate line, size: $48 \times 59\,\text{mm}$. ©2009 IEEE; reprinted with permission from [179].

3.2.5 Multiple resonance antennas

Multiple resonance antennas are combinations of multiple, narrowband, radiating elements. Each element, for example a dipole, covers a limited bandwidth, e.g. 20% of the total UWB bandwidth. Typical candidates are the logarithmic periodic (log-per) dipole arrays [128] and fractal antennas.

The planar log-per antenna (see example in Fig. 3.15) consists of n adjoining unit cells (dipoles), with the dimensions l_μ of adjacent cells scaled by $\log(l_\mu / l_{\mu+1}) = \text{constant}$ [131]. Each dipole is etched with one half on the top layer and the other half on the bottom layer of the substrate. The antenna is fed, for example, by a coaxial line via a triplate line inside the structure at the high frequency port. This structure can be optimized for

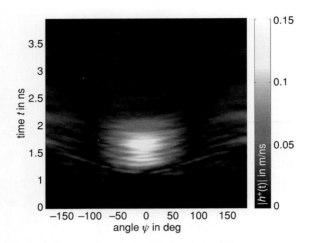

Figure 3.16 Measured impulse response $\left|h^+\left(t,\psi\right)\right|$, $\theta = 90°$ of the log-per antenna (Fig. 3.15) in the E-plane. ©2009 IEEE; reprinted with permission from [179].

low return loss ($S_{11} < -10$ dB) in the whole FCC UWB frequency band. The design of the antenna is compact ($60 \times 50 \times 2$mm), with 3 dB beamwidths $\psi_{3dB} \approx 65°$ in the E-plane and $\Theta_{3dB} \approx 110°$ in the H-plane. These values are quite constant over the desired frequency range. For reasons of comparison with other antennas, Fig. 3.16 shows the transient response $h(t, \psi, \theta = 0)$ versus the E-plane angle ψ and time t. The broadening of the impulse response compared to the Vivaldi antenna is obvious. These strong oscillations can be explained by the consecutively excited ringing of coupled, resonating dipoles. Consequently the peak value p of the antenna impulse response reduces to only 0.13 m/ns due to the resonant structure of the radiating dipoles. Any UWB antenna with resonant elements broadens the radiated impulse, i.e. increases the τ_{FWHM} and lowers the peak value p.

Figure 3.17 is a sectional view of the main beam of the transient response $h(t, \psi = 0°, \theta = 90°)$. The resonant character of the log-per antenna is even more obvious in this representation, where the characteristics are determined to $\tau_{FWHM} = 805$ ps and $\tau_{r=0.22} = 605$ ps. Fig. 3.18 shows the transfer function $H(f, \theta)$ of the log-per antenna. The antenna exhibits a relatively constant and stable radiation pattern over the frequency range. The collapse of the transfer function in the main beam direction at particular frequencies, and simultaneous side radiation at the same frequencies are noticeable. This is due to the excitation of higher order modes, e.g. λ-resonances, of the "$\lambda/2$-dipoles". The main parameters of the log-per antenna of Fig. 3.15 are shown in Table 3.2.

3.2.6   Electrically small antennas

Electrically small antennas [185] are "equally bad" for any desired UWB operation concerning impedance matching and radiation. These antennas are far below

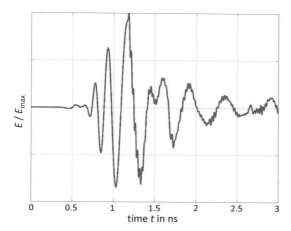

Figure 3.17 Simulated impulse response of the log-per antenna (Fig. 3.15) in the main beam direction. ©2009 IEEE; reprinted with permission from [179].

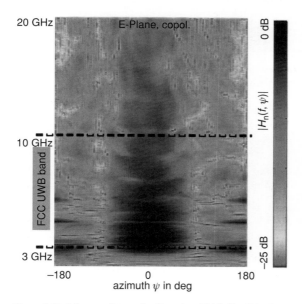

Figure 3.18 Measured transfer function $H(f, \theta)$ of the log-per antenna vs frequency in the E-plane. ©2009 IEEE; reprinted with permission from [179].

resonance, i.e.

$$a < \frac{\lambda}{10}, \tag{3.26}$$

where a is the antenna dimension. Thus similar conditions for all frequencies exist. With a proper impedance transformation the antennas can be made UWB. Typical candidates are the different types of D-dot probe antennas, small monopole antennas [63] and the Hertzian dipole.

Table 3.2 UWB parameters of the log-per antenna of Fig. 3.16 in the main beam direction.

Parameter	Value
p_{max} (m/ns)	0.13
τ_{FWHM} (ps)	805
$\bar{G}$ (dBi)	4.5
G_{max} (dBi)	6.8
$\tau_{r=0.22}$ (ps)	605

Figure 3.19 Monocone antenna with enlarged ground plane, with $d = 80$ mm diameter. ©2009 IEEE; reprinted with permission from [179].

Figure 3.20 Planar monopole antennas with CPW and microstrip feed.

In Fig. 3.19 a typical rotational symmetric UWB monocone antenna is shown. The monocone antenna, an asymmetric structure, does not require any balun for an asymmetric feed line, however (in theory) it does need an infinite ground plane, which is cut for practical applications. For the bandwidth under consideration, monocone antennas with a ground plane diameter larger than 40 mm exhibit a return loss below $S_{11} = -10$ dB. The finite ground plane also affects the stability of the radiation pattern versus frequency and the impulse radiating properties. In [26], an antenna has been proposed that overcomes this problem, compensating the dominating antenna capacity of electrically small antennas by inductive coupling to sectorial loops. The monocone properties can be well approximated by planar structures such as planar monopoles (see Fig. 3.20). These are very well suited for short-range communications as they can easily

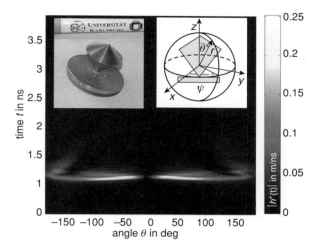

Figure 3.21 Monocone antenna and its measured impulse response $|h^+(t)|$ in the E-plane (ground plane diameter $d = 40$ mm). ©2009 IEEE; reprinted with permission from [179].

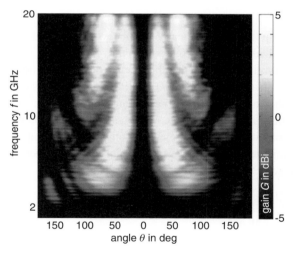

Figure 3.22 Measured gain $G(f, \theta)$ of monocone antenna vs frequency in the E-plane. ©2009 IEEE; reprinted with permission from [179].

be integrated in different planar lines and circuits. The monocone antenna has an omni-directional radiation pattern in the H-plane. The impulse response $h(t)$ and gain $G(f)$ for a monocone antenna with reduced ground plane ($d = 40$ mm) are shown in the E-plane in Figs. 3.21 and 3.22 respectively. It can be seen that the impulse response is short, which indicates small ringing, and the antenna radiates over a wide elevation angle θ where $10° < \theta < 90°$ with a relatively constant gain $G(f)$. At higher frequencies the radiation is more upwards and a second beam emerges from the ground plane. Because of the omnidirectional character of the antenna, the small $\tau_{\text{FWHM}} = 75$ ps value

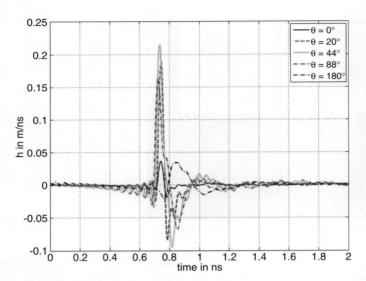

Figure 3.23 Measured impulse response $h(t, \theta, \psi = 90°)$ as a function of elevation θ for the monocone antenna. ©2009 IEEE; reprinted with permission from [179].

of the antenna impulse response, and the nearly frequency-independent gain, the mono-cone antenna is often applied for channel measurements.

In narrowband operations it is assumed that the antenna radiates identical signals in all directions of the antenna characteristic $C(\theta, \psi)$. In ultra-wideband this cannot be taken as granted, as will be shown for the monocone antenna. Fig. 3.23 shows the impulse response of the monocone antenna for $\psi = $ const. versus the elevation angle θ. It can clearly be seen that the radiated signals are elevation angle θ dependent. In a multipath environment these signals overlap at the receiver, which may cause severe distortion. Proper channel models can be used to study these effects [110].

3.3   UWB antenna system aspects

In practice, from a system point of view, two cases for UWB have to be distinguished:

- multiple narrow bands, e.g. OFDM (ECMA-368 Standard)
- pulsed operations (IEEE 802.15.4a).

The first case can usually be treated similarly to the well-known narrow band operations. The relevant criteria are well covered by the frequency-dependent transfer function $\mathbf{H}(f, \theta, \psi)$. Antennas for these applications could be any of the types already discussed, but especially the log-per antenna.

The second case needs a closer look. If in a pulsed operation for radar or com-munications the full FCC bandwidth from 3.1 to 10.6 GHz (i.e. 7.5 GHz) is covered,

peak value p in m/ns	0.35	0.13	0.10	0.13	0.23
τ_{FWHM} in ps	135	140	290	805	75
$\tau_{r=0.22}$ in ps	150	185	850	605	130

Figure 3.24 Characteristic parameters of the presented UWB antennas. ©2009 IEEE; reprinted with permission from [179].

for example, with the derivative of the Gaussian pulse with $\tau_{FWHM} = 88$ ps, then the transient behavior, the impulse response $\mathbf{h}(t, \theta, \psi)$ of the antenna, has to be taken into account. In this case the impulse distortion in the time domain and in the spatial domain have to be examined for compatibility. An adverse behavior of the impulse response $\mathbf{h}(t, \theta, \psi)$ with the following problems:

- low peak magnitude $p(\theta, \psi)$
- very wide pulse width τ_{FWHM}
- long ringing τ_r

influences the system characteristics, for example:

- the received signal strength $u_{Rx}(t)$, SNR
- the data rate in communications
- the resolution in radar.

These adverse effects set requirements for the antennas, but also for the other UWB hardware front-end elements like amplifiers, filters, equalizers, detectors, and so on. These requirements restrict the potential antennas to small antennas or traveling-wave antennas, such as monocone, bow-tie, Vivaldi, and horn. All antennas with resonances or spurious surface currents are bad candidates and should be disregarded – among them is definitely the log-per antenna.

For certain cases where circular polarization is required, further restrictions hold. A logarithmic spiral antenna can, for example, only radiate circular polarization if the pulse duration is longer than the equivalent circumference of the active radiating zone. For 88 ps pulses, this should be less than 2.6 cm. These statements show that in addition to research into UWB at the component level, research at the system level has to be performed. Ultra-wideband as an emerging technology requires a thorough knowledge of the antenna behavior in the time domain, the frequency domain and, in certain cases, the spatial domain. It has been shown that for ultra-wideband, certain antenna classes can be defined according to their radiating characteristics. In Fig. 3.24, typical, relevant data of the discussed UWB antennas are compared.

3.4   Polarization diversity antennas

Nearly every wireless application can profit from an extension of its performance by polarization diversity. In the radar/imaging, more extensive information about the target is provided due to the different scattering behavior of most objects for different polarizations. A higher robustness of the system is provided in localization applications, since during propagation the signal polarization may change, leading to poor or even failed signal detection. In communication systems the channel capacity and system robustness may be significantly improved as well. Multiple Input Multiple Output (MIMO) systems especially profit from an application of dual-polarized antennas. This section describes the specific requirements on UWB polarization diversity antennas and gives examples of practical realization of such radiators for different applications.

As previously mentioned, a signal may be radiated in different polarizations, e.g. elliptical, circular and linear [24]. Since the linear polarization is provided by most types of radiators, we will only consider this type of polarization here. A polarization diversity means a radiation capability of two orthogonal polarized signals independent of each other. In the case of linear polarization, the radiated electric fields of both signals are always oriented 90° to the propagation direction and at 90° to each other. In the following, the term *dual-polarized antennas* is used (the shortened version of dual-orthogonal-polarized antennas).

3.4.1   Requirements for UWB polarization diversity antennas

In the previous sections, the main specific parameters describing a radiator suitable for UWB systems have been pointed out: sufficient impedance matching in the desired bandwidth, constant main beam direction(s), possibly stable beamwidth over frequency, and constant position of the radiation phase center over frequency. The last property is equivalent to linear phase response, i.e. constant group delay in the given direction over frequency. In the case of polarization diversity, the additional requirements for antennas shall be fulfilled for a proper functionality. These are as follows.

- *Small coupling between the ports for orthogonal polarizations.* No energy should be transferred from the port which radiates the desired polarization to the port which radiates the orthogonal one. In the case of absorptive termination of the port for orthogonal polarization, a high coupling results only in a gain loss, but in the case of reflective termination a proportional part of the energy is radiated in the orthogonal polarization and resonances occur. A typical value for decoupling between the ports is >20 dB, within the desired bandwidth.
- *High polarization purity.* This value describes the ratio between the energy radiated in the desired polarization to the energy contained in the orthogonal one. It is expected that the power fed into the antenna port is mostly radiated in the desired polarization and only a negligible part of it is oriented orthogonally to the intended one. The requirement on the polarization purity (or far-field polarization decoupling) depends on the application and may vary from, say, 10–15 dB in simple communication or

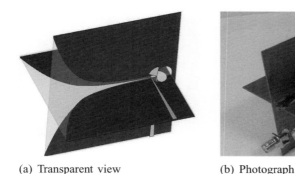

(a) Transparent view   (b) Photograph

Figure 3.25 Dual-polarized, tapered slot antenna. ©2008 MIKON; reprinted with permission from [11].

localization systems to 35–40 dB or higher in high-end radar applications. A typical value is 20 dB.

- *Same position of the radiation phase centers for both polarizations.* It is expected that the signal is radiated in both polarizations physically from the same point. An offset between the radiation points of both polarizations implies a priori different propagation conditions, which cannot be always compensated.
- *Similar beam widths in orthogonal planes (e.g. E- and H-planes).* This requirement is applied only to the directive antennas and is valid only for special applications. In some applications it is desirable that the same part of space is illuminated independent of the polarization. An example of the importance of this property is an imaging system, which may deliver different resolutions in orthogonal planes if antenna beamwidths in the respective planes differ.

In the next part, some practical examples of UWB polarization diversity antennas are described and characterized.

3.4.2   Design example 1: dual-polarized traveling wave antennas

As described in the previous sections, traveling wave antennas are very good candidates for pulse radiating, directive antennas. The tapered slot antenna, also known as the Vivaldi antenna, exhibits a relatively high gain, constant main beam direction and low pulse distortion properties [179]. These advantageous properties can be transfered to a dual-polarized version. An example of such a radiator is shown in Fig. 3.25 [11]. In order to obtain a dual-orthogonal-linear polarization, two radiators have to be crossed orthogonally. It causes a necessity of intersection of both antennas, where the width of the feeding slot of each antenna must be larger than the thickness of the substrate used as a carrier for the metalization. This implies a high antenna input impedance, which introduces a challenge in the design of the antenna feed. For that reason, special attention should be given to the choice of substrate. In the example shown in Fig. 3.25, an aperture coupling is used. However, direct feeding techniques – such as in the case

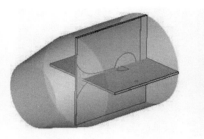
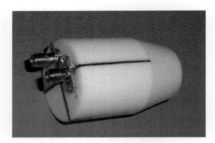

(a) Transparent view   (b) Photograph

Figure 3.26 Dual-polarized dielectric rod antenna with crossed tapered slot feed. ©2010 IEEE; reprinted with permission from [12].

of antipodal Vivaldi antennas (e.g. [86]) – can also be used. In this case, a symmetric excitation of the slot introduces a problem due to the large slot width. It might result in an asymmetrical, frequency-dependent radiation pattern in the E-plane of the respective polarization. The antenna in Fig. 3.25 has the same phase center of radiation for both polarizations and relatively similar beamwidths in both planes. The intersection of the two structures introduces a negligible additional pulse distortion, thus the antenna can be applied successfully in pulse-based UWB systems. A mean polarization decoupling better than 20 dB offers sufficient performance for many applications where polarization diversity is desired.

The dual-polarized tapered slot antenna in Fig. 3.25 has a height of 62 mm. If the antenna is to be used in an array, this dimension defines the minimum distance between the array elements. If the distance is large in comparison to the wavelength, gratings lobes arise and move towards lower frequencies with increasing distances (see Chapter 4). In order to minimize this effect, the distance between the array elements, i.e. the transversal dimensions of the antenna, need to be reduced. The possibility of reducing the antenna dimension introduces an embedding of the antenna in a dielectric. An example of such realization is shown in Fig. 3.26. The desired radiation properties, such as wideband behavior and low distortions, can be preserved only for relatively small electric permittivities of an embedding material. On the contrary, a high permittivity is needed for a significant reduction of the dimensions. Hence minimization of the antenna dimensions has its limitations. In practice, the relative dielectric permittivity (ε_r) should be no higher than ≈ 3.5–4. With this approach, an antenna diameter of 35 mm is achieved with a sufficient impedance match between 3.1 and 10.6 GHz [12]. An application of dielectrics also gives an additional degree of freedom in the shaping of the radiation pattern, which is influenced by the shape of the material. In the presented solution the dielectric cone is cut off in order to avoid side lobes at higher frequencies [12]. The antenna exhibits a directive radiation pattern with constant main beam direction versus frequency, low pulse distortion and relatively high polarization decoupling. Therefore it also suits IR-UWB systems. An additional advantage of embedding in a dielectric is a significantly improved mechanical stability of the device.

(a) Top   (b) Bottom

Figure 3.27 Photographs of the dual-polarized differentially fed 4-slot antenna. ©2009 EurAPP; reprinted with permission from [9].

3.4.3   Design example 2: dual-polarized antennas with self-cancellation of cross-polarizations

The previously described antennas can be realized in the planar technique; however, due to realization requirements, all three dimensions are needed for the radiation of both orthogonal polarizations. For many applications a fully planar solution is desired. This automatically implies a bi-directional radiation pattern, which is in general oriented perpendicularly to the antenna surface. If an antenna with a single beam is desired, the second beam needs to be either reflected by a wideband reflector or absorbed. In the following, several solutions for the planar realization of dual-polarized UWB radiators are shown. They are based on dual feeding principles for the radiation of single polarization, i.e. four feeding points for dual-polarized antenna. The increased feed complexity is compensated by a superior performance regarding port decoupling, polarization purity and position of the phase centers of radiations, as shown in the examples below.

A practical realization of a planar antenna with self-cancellation of cross-polarization is shown in Fig. 3.27. The antenna consists of four tapered slots connected to each other as shown in the photograph. The antenna therefore has four feeding ports, whereas two opposite ones would need to be fed simultaneously for the radiation of single linear polarization. The principle of the antenna is explained by looking at Fig. 3.28(a), where the schematic electric field distribution in the antenna is shown. For the radiation of the vertical polarization the antenna has to be fed at ports 1.1 and 1.2. The signal at these ports must be equal in amplitude and 180° out-of-phase, i.e. the phase difference is 180° in the whole designated frequency range. Due to the symmetrical orientation of the feeding lines, the electric field vectors in the respective feeding slots are in-phase and so interfere with each other in the middle of the structure. The distribution of the electric fields into co- and cross-polarized components (see Fig. 3.28(b)) shows that the co-polarized components that originated from opposite slots are in-phase, i.e. the vectors are in the same direction. On the contrary, the cross-polarized electric field vectors have opposite directions, i.e. are out-of-phase. For the far-field antenna properties it means that the co-polarization is enhanced during radiation, whereas the cross-polarization is

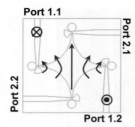
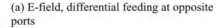
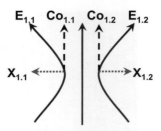

(a) E-field, differential feeding at opposite   (b) Schematical E-field distribution
ports

Figure 3.28 Principle of the dual-polarized differentially fed 4-slot antenna. ©2009 EurAPP; reprinted with permission from [9].

suppressed. The suppression of cross-polarization is dependent mainly on the amplitude and phase imbalance of the feeding signals at opposite ports. If these signals are equal in amplitude and phase, a total cancellation of cross-polarization in the E- and H-planes for the main beam direction (of the co-polarization) is achieved. In practice, a polarization purity of more than 25–30 dB can be achieved.

In Fig. 3.28(a) it can be observed that the electric field produced by the in-phase feeding of the opposite slots is unable to couple to the slots for orthogonal polarization. It results in a very high decoupling of the ports for orthogonal polarizations, which is a positive spin-off of the feeding technique. A decoupling larger than 30–40 dB can easily be achieved in the practical case. Another advantage of the solution is high stability of the phase center of radiation. Due to symmetrical feeding of the antenna, the phase center is positioned directly in the middle of the structure and is constant over the frequency. Since the structure geometry is identical for both orthogonal polarization planes, it is evident that the phase center of radiation is exactly the same for both polarizations.

It can be concluded that the solution assures wideband radiation with a very stable location of the phase center of radiation and its position is the same for both polarizations. The coupling between the ports for orthogonal polarizations is reduced due to the orthogonal placement of the feeds. The antenna is suitable for the realization in a planar technique and can be integrated into any device where a sufficiently large area of metallic plane is available. The consequence of planar realization is a bi-directional radiation pattern with main beam directions perpendicular to the antenna surface. Due to similar dimensions of the aperture in all directions, the beamwidths in all planes are comparable. The simulations show that this specific example can be applied to UWB systems with a relative bandwidth up to $\approx$35%, i.e. the ECC UWB frequency mask from 6 to 8.5 GHz can be covered. For larger bandwidths, e.g. the FCC UWB frequency mask from 3.1 to 10.6 GHz, another solution can be used.

Relative bandwidths of over 100% can be achieved with the solution shown in Fig. 3.29(a). The antenna consists of four elliptical monopoles surrounded by a ground plane. The geometry of the ground plane may be shaped to many forms; however, in

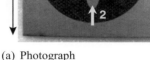
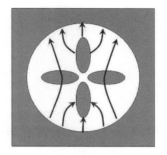

(a) Photograph

(b) Schematical E-field
distribution

Figure 3.29 Principle of dual-polarized differentially fed 4-ellipse antenna. ©2009 IEEE;
reprinted with permission from [4].

order to assure the maximum possible symmetry of the radiation patterns, a circular
shape will be chosen [8]. Two inline monopoles are excited simultaneously for the
radiation of single linear polarization. The monopoles shall be placed symmetrically
w.r.t. the middle of the structure in order to achieve maximum symmetry of the electric
field distribution. The two remaining monopoles are dedicated for the radiation of the
orthogonal polarization and are therefore placed orthogonally w.r.t. the other pair. The
monopoles are fed between their tips and the ground plane. This allows for a spatial
separation of the antenna feeds, which simplifies the design. The orientation of the electric field at the excitation points for the radiation of single polarization is marked by the
arrows in Fig. 3.29(a). The orientation of these vectors in the global coordinate system is
co-aligned. However the feeding of both monopoles with respect to the antenna ground
plane is differential.

Each of the monopoles is able to radiate separately over a large bandwidth. If the
remaining three monopoles are removed from the structure, a so-called volcano smoke
[166] or diamond-shaped antenna [165] evolves. Such an antenna exhibits two main
beam directions which are symmetrical to the antenna surface. The beams change their
directions in the E-plane depending on the frequency, which results in non-symmetrical
antenna geometry in the respective (E-)plane. The radiated electric field shows a very
high cross-polarization level which is also caused by the antenna geometry. These
drawbacks are significantly suppressed by the addition of a second monopole and differential excitation of the pair. A schematic electric field distribution during differential
excitation of opposite monopoles is shown in Fig. 3.29(b). It is evident that the field
distribution is symmetrical w.r.t. the axis crossing the feeding points. A similar distribution of the fields into co- and cross-polarized components, as in Fig. 3.28(b), results
in the enhancement of the co-polarized and suppression of the cross-polarized components. The effectiveness of the suppression of cross-polarization is dependent on the
radiation angle. In order to preserve this desired property over a wide angular width,
the distance between the monopoles and feeding point should be kept small w.r.t. the
wavelength [7].

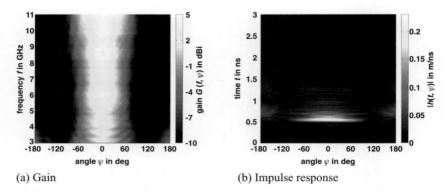

(a) Gain   (b) Impulse response

Figure 3.30 Measured characteristics of dual-polarized differentially fed 4-ellipse antenna in the H-plane, co-polarization. (a) ©2009 IEEE; reprinted with permission from [4]; (b) ©2010 KIT Scientific Publishing; reprinted with permission from [3].

The symmetrization of the electric fields in the radiation zone results in a stable phase center of radiation over frequency, which is placed directly in the middle of the structure between the tips of the monopoles. Due to rotational symmetry of the antenna geometry, the phase center of the orthogonal polarization is at exactly the same place. Similarly to the case of the previously described four-slot antenna, the antenna geometry and feeding technique prevent the coupling of the energy to the ports for orthogonal polarization. This effect simplifies the design, since no additional decoupling techniques need to be applied.

The antenna in Fig. 3.29 has dimensions of $40 \times 40\,\text{mm}$, with a diameter of the opening in the ground plane of 31.6 mm. The antenna is optimized for the frequency range 3.1–10.6 GHz. It means that the circumference of the ground plane opening is nearly equal to the wavelength at the lowest frequency, and introduces a practical limit for the lower cut-off frequency. The optimization parameters are the shape and dimensions of the monopoles, which have an influence not only on impedance matching, but also on the radiation pattern.

The antenna maintains two main beam directions oriented perpendicularly to the antenna surface. In order to obtain a monodirectional radiation, either a wideband reflector must be applied, or one of the beams must be absorbed. The first solution increases the gain of a single beam by approximately 3 dB, however it introduces a difficulty to the reflector design which must be able to reflect the wave with the same phase behavior over the whole designated frequency range. The second solution suffers from a reduction of total antenna efficiency of approximately 3 dB. A measured gain for the H-plane in co-polarization over the frequency is shown in Fig. 3.30(a). Although the antenna has been equipped with an absorber on one side of the surface, the maximum achievable gain is in the range of 5 dBi. The previously described principles regarding the constancy of the main beam and symmetrical pattern are preserved. The beam width is relatively constant over the frequency. Also evident is the lack of side lobes, which is an effect of a simple antenna structure. From the antenna impulse response in Fig. 3.30(b),

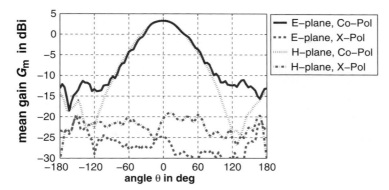

Figure 3.31 Measured mean gain G_m of dual-polarized differentially fed 4-ellipse antenna. ©2010 KIT Scientific Publishing; reprinted with permission from [3].

it can be seen that the pulse delay for different angles is nearly constant; this is an effect of the geometrical constancy of the phase center of radiation. This property is especially advantageous for localization or radar/imaging systems, since no correction of the pulse delay for different angles needs to be performed. Furthermore, it can be observed that the impulse response is very short and exhibits short ringing. It can be concluded, according to the previous chapter, that the phase response is linear over frequency and the group delay is constant. It means that all frequencies are radiated coherently with minimum distortion. All these properties show that the radiator is well suitable for pulse-based UWB systems with relative bandwidths exceeding 100%.

Figure 3.31 shows the mean gain according to (2.26) for the four-ellipse antenna for the E- and H-planes, co- and cross-polarization. The evaluated frequency range is from 3.1 to 10.6 GHz. Firstly, it can be noted that both planes have the same main beam direction, which is symmetrical around 0°. Secondly, it can be seen that both planes maintain nearly the same beamwidths, which is a result of similar effective aperture dimensions in the respective planes. Consequently it is assured, due to the design, that the same part of space is illuminated in both polarizations. This is especially advantageous in imaging applications, as shown in Section 6.4. Since the antenna is intended for applications in polarization diversity systems, the polarization purity is vital. As can be seen in the diagram, the mean polarization decoupling in the main beam direction is above 20 dB for both planes. The cross-polarization in this case is mainly caused by nonidealities in the feeding network, like phase and amplitude imbalance between the signals feeding opposite monopoles. The radiation from the feeding network itself is also a reason for radiation in cross-polarization [4].

Directive antennas are especially interesting for radar or point-to-point communication. However for mobile terminals in localization or communication systems, omnidirectional antennas are of great interest. Fig. 3.32(a) shows a nearly omnidirectional solution for a dual-polarized UWB antenna. It consists of two crossed pairs of diamond-shaped dipoles. The antenna is fed in the middle of the structure with a separate coaxial cable for a single flare of the dipole. In order to radiate single polarization, two opposite

(a) Photograph   (b) Antenna feeding

Figure 3.32 Omni-directional (in respective H-plane), dual-polarized, differentially fed antenna. ©2010 EuMA; reprinted with permission from [14].

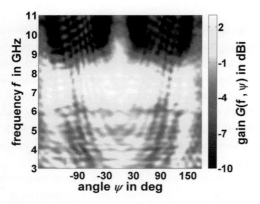

Figure 3.33 Measured gain of an omni-directional, dual-polarized, differentially fed antenna in the H-plane: co-polarization. ©2010 EuMA; reprinted with permission from [14].

flares (i.e. one dipole) have to be excited with differential signals. The inner conductors of the coaxial cables are connected to the tips of the respective flares, whereas the outer conductors are placed in close proximity to each other as shown in Fig. 3.32(b). The signal, after termination of the outer conductors, is transmitted between the metalizations with the highest potential differences, i.e. between both inner conductors. They are connected to the tips of the dipole, and realize a possibility of dual-polarized excitation of the antenna with high decoupling between the ports.

The antenna maintains a high polarization decoupling, similar to the example of differentially-fed four-ellipse antenna, but with an omnidirectional radiation pattern in the H-plane. In the E-plane the radiation characteristic is bi-directional, hence the overall antenna pattern is dipole, the same as for a very wide frequency range. The measured gain of the antenna, designed for the low cut-off frequency of 3.1 GHz, is shown in Fig. 3.33. Note that the omnidirectional characteristic is achieved up to the frequency of approximately 9 GHz. Above this frequency the antenna radiates like a slot antenna with four directions defined by the four slots between the metallic flares. This

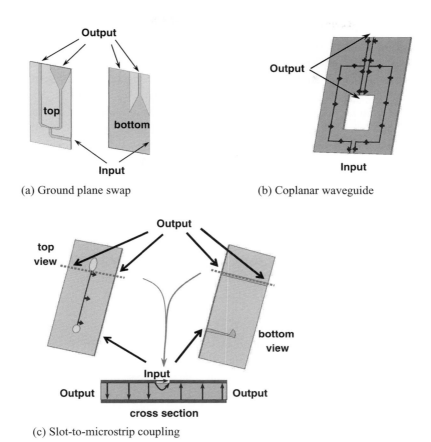

(a) Ground plane swap

(b) Coplanar waveguide

(c) Slot-to-microstrip coupling

Figure 3.34 Principles for frequency-independent generation of differential signals.

introduces practical limitations to the useful bandwidth of this antenna. However the omnidirectional pattern is preserved for a relative bandwidth of approximately 100%, which still represents a sufficient bandwidth for IR-UWB systems. Due to the realization in the planar technique, this particular antenna is a good candidate for integration into small, mobile devices.

3.4.4 Frequency-independent 180° power splitter

The prerequisite for the radiation of pure linear polarization with the antennas presented in previous sections is feeding with two 180° out-of-phase signals. It means that the pulses fed to the antenna have to be equal in amplitude and shifted by 180° over the whole desired frequency range w.r.t. each other. There are many methods for the generation of such signals.

A trivial solution is based on a 2-way (3 dB) microstrip power divider and is reached by swapping the microstrip line with the ground plane in one of the branches. An example is shown in Fig. 3.34(a). Theoretically, the signals at the outputs have the same amplitude and due to the swap of the ground plane and microstrip, a 180° phase shift between the output signals is achieved. In practice some reflections occur at the transition from microstrip to ground plane. A very low phase imbalance results, however a small amplitude ripple at both outputs over the frequency is still present. This leads to amplitude imbalance, which decreases the performance of the radiator. A second method is shown in Fig. 3.34(b). In this case a coplanar waveguide structure is used. The line is split into two independent slotlines, which are combined again to a coplanar waveguide to build a 3 dB power combiner. In this case a very good 180° phase imbalance is also achieved and, due to the symmetrical arrangement of the structure, the amplitude imbalance is very small too. However the transition from coplanar waveguide to slotlines and back leads to increased radiation from the structure, which decreases the efficiency.

Another solution incorporates a transition from slotline to microstrip line with a 3 dB power divider. The principle is shown in Fig. 3.34(c). Firstly a transition from microstrip line to slotline is applied. In the next step the signal is coupled to the microstrip line via aperture coupling. The signal is guided in the microstrip line symmetrically in both directions to the outputs of the feeding network. Due to the symmetrical arrangement of the structure, the amplitude and phase imbalance between the signals at both outputs is very low. The signals at the outputs are 180° out-of-phase due to the electric field arrangement at the transition from slotline to microstrip line (see bottom of Fig. 3.34(c)). If the electric field is in the opposite direction, it means there is a 180° phase shift. Since no length-dependent elements are applied in the structure, the out-of-phase signal distribution is frequency-independent and can be successfully applied in UWB systems. The useful operational bandwidth is limited by the bandwidth of the slotline to microstrip transition. In practice, relative bandwidths of more than 150% may be achieved.

3.5   UWB antennas for medical applications

Lately, there has been an increase of interest in scientific research in UWB techniques for medical applications. By combining UWB and radar, new medical sensors with very high resolution are feasible. Different from the X-ray based imaging and tomography, UWB radar probes use non-ionizing electromagnetic waves, making the treatment less harmful for the patients. Compared to ultrasound, which is unable to penetrate bones and air inside the body, and causes acoustic shadows [84], [104], UWB signals provide the potential for imaging of the lung or the brain due to their good penetration capability [160].

Besides medical diagnostics, research around the world is currently also focusing on the potential benefits of employing UWB for wireless communications of medical sensors (i.e. implants) and external devices. The UWB technique is very attractive for Wireless Body Area Networks (WBAN) and in-body communications due to its achievable high date rate [49, 60]. Additionally, the low complexity of UWB systems makes the potential sensors cheap and hence widely available.

However, the antenna design for a medical application differs from that for free-space operation. Due to the presence of complex human body medium, the impedance matching and bandwidth of the antenna are strongly influenced. The antennas suffer from reduced efficiency, radiation pattern fragmentation and variations in feed-point impedance. In the case of on-body matched antennas, where the antennas are placed directly on the human body, the radiation into multiple layers of a lossy medium is considered. The design procedure becomes more complicated than for simple free-space operation scenarios, but the power radiated into the body increases drastically compared to an antenna placed at some distance to the body. Furthermore, many features like small size, low weight and low cost are required for medical applications.

The challenges in the antenna design for a UWB medical system can be summarized as follows:

- human tissue is a dispersive medium with frequency-dependent dielectric properties
- human bodies are complicated and unique
- the antenna must not be sensitive to the dielectric properties of human tissue
- the antenna should be optimized for operation directly on or near to the skin
- the radiation property should not change significantly in the whole frequency band
- the size of the antenna must be minimized to fit to the human body.

Verification of the antenna by measurements becomes rather complicated. A tissue-mimicking medium is required to emulate the human tissue around or in front of the antenna. The measurements have to be performed in the near-field range of the antenna due to the high signal attenuation of the medium. Details will be discussed further in Section 3.5.3.

3.5.1   Analysis of the dielectric properties of human tissue

The dielectric properties of human tissue should be studied before starting to discuss UWB antennas for medical applications. The human body is modeled as a multilayer structure consisting of different tissues such as skin, fat, muscle and bone, which possess different dielectric properties [52]. It is important to underline that the dielectric layers of tissues are lossy mediums. In the model to be considered, effects such as attenuation and frequency dispersion in the whole UWB band cannot be neglected. Considering the dielectric loss and conductive (or Joule) loss, the complex permittivity of a lossy medium can be written as follows [190]:

$$\varepsilon(\omega) = \varepsilon_0 \left[\varepsilon_r'(\omega) + j\varepsilon_r''(\omega) \right] = \varepsilon'(\omega) - j\frac{\sigma}{\omega} \tag{3.27}$$

where $\varepsilon_r'(\omega)$, $\varepsilon_r''(\omega)$ and σ are the real and imaginary relative permittivities and the conductivity of the medium, respectively. ε_0 is the permittivity of free space and ω is the angular frequency. The conductive loss, i.e. the dominant loss of EM waves in human tissue, is described in the equation by the imaginary part. The polarization loss is not considered. The wave equations are solved including the complex permittivity with conductive loss, which yields the complex propagation constant

$$\gamma = \alpha + j\beta = j\omega\sqrt{\mu\varepsilon' - j\frac{\sigma}{\omega}}. \tag{3.28}$$

The magnetization loss is not considered either. The real and imaginary parts of the complex propagation constant are:

$$\alpha = \frac{\omega}{c_0} \sqrt{\frac{\varepsilon_r'}{2} \left[\sqrt{1 + \left(\frac{\varepsilon_r''}{\varepsilon_r'}\right)^2} - 1 \right]} \qquad (3.29)$$

$$\beta = \frac{\omega}{c_0} \sqrt{\frac{\varepsilon_r'}{2} \left[\sqrt{1 + \left(\frac{\varepsilon_r''}{\varepsilon_r'}\right)^2} + 1 \right]}. \qquad (3.30)$$

The attenuation coefficient α and phase coefficient β can be determined by a known complex permittivity. According to the Cole–Cole dispersion, the complex relative permittivity $\underline{\varepsilon}$ of human tissue is described in (3.31) [53, 54], in which α_n is the distribution parameter, τ is the relaxation time, $\Delta\varepsilon$ denotes the magnitude of the dispersion, σ_i is the static ionic conductivity:

$$\varepsilon(\omega) = \varepsilon_\infty + \sum_n \frac{\Delta\varepsilon_n}{1 + (j\omega\tau_n)^{(1-\alpha_n)}} + \frac{\sigma_i}{j\omega\varepsilon_0}. \qquad (3.31)$$

The frequency-dependent relative permittivity from 1 to 10 GHz of different tissues and the attenuation coefficient in the tissues are shown in Fig. 3.35. Distilled water, as a reference, has a high relative permittivity of approximately 80 at 2 GHz. The relative permittivities of various tissues are strongly dependent on their water content. The relative permittivity of fat is around 5.5, while skin and muscle have a large relative permittivity of around 45 and 55 at 2 GHz, respectively. The signal attenuation in the respective materials are shown in Fig. 3.35(b). Fat has a small attenuation coefficient, while water, skin and muscle have a relatively high one and are frequency-dependent. Hence a rather high attenuation and distortion of EM signals in these tissues is expected, but first investigations predict that the dynamic range of a highly sophisticated UWB radar will still be sufficient [92].

3.5.2   UWB on-body antennas

The relative permittivity of skin is >40. When reflected at the air–skin interface more than 70% of the energy of an EM wave is scattered back. Therefore, antennas for medical diagnostics must be matched to the skin. However, a different wavelength and wave impedance have to be considered if the antenna is to be placed close to or directly on the skin. In the design procedure, the antenna is optimized together with a phantom model of tissues. A simple model based on flat layers is shown in Fig. 3.36. Typical thickness values of each tissue layer are based on the investigations on adults (aged 20–60) in [35]. Based on the analysis on signal attenuation in tissues presented in the previous section, a low frequency range would preferably be chosen for UWB medical diagnostics. On the other hand the antenna dimensions will increase when the frequency is decreased. Besides a low attenuation, a high bandwidth is still required for a good measurement resolution. Therefore the frequency range of 1–9 GHz seems to be a good compromise for most medical diagnostics applications. In the following

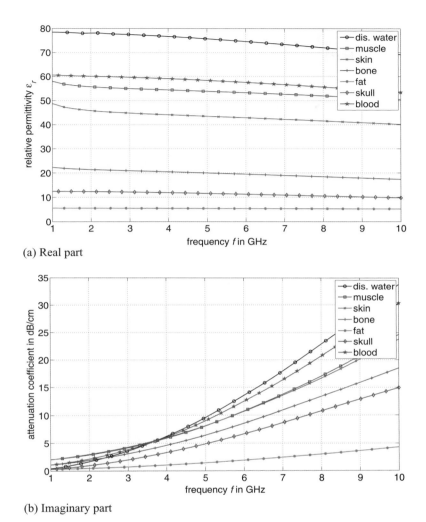

(a) Real part

(b) Imaginary part

Figure 3.35 Relative permittivity and the attenuation coefficient in human tissue (distilled water, muscle, skin, bone, fat, skull and blood) over frequency (distilled water: measured by the author as a reference; other tissues: predicted by the Cole–Cole model). ©2012 IEEE; reprinted with permission from [94].

example, the design of a body-matched antenna for this frequency range is presented. The major design goal is that the antenna should be as small as possible to enable usage in an antenna array, which fits on the part of the body to be investigated (head, breast, etc.).

The antenna design is based on the concept in [4], which was characterized for free-space propagation. The designed antenna is now matched to human skin to avoid a strong reflection of the incident UWB pulse at the skin boundary. The layout of the antenna structure and the feed network are shown in Fig. 3.37. The antenna is fabricated on the substrate Rogers RT 6010 ($\varepsilon = 10.2$, $d = 1.27$ mm, $\tan \delta = 0.0023$). Two monopoles

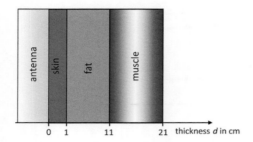

Figure 3.36 Geometry of tissue layers for the design of on-body antenna.

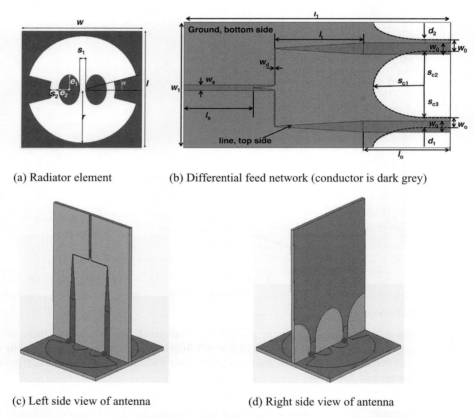

(a) Radiator element

(b) Differential feed network (conductor is dark grey)

(c) Left side view of antenna

(d) Right side view of antenna

Figure 3.37 Layout (radiator element and differential feed network) and 3D views of the antenna (electric connection between radiator and feed network marked with circles).

and a sector-like slot are on the top of the substrate. The symmetrical arrangement of the two monopoles in the circle helps to maintain a symmetrical current distribution in the radiation zone around the center of the antenna. The phase center of the radiation lies exactly between the two monopoles in the center of the structure at all frequencies, due to the symmetry of the current distribution [4].

Table 3.3 Design parameters of the antenna and differential feed network [94].

Parameter	w	l	s_1	s_2	e_1	e_2	r	α	w_1	l_1	w_s
Value (mm)	35	35	1.8	2.2	4.5	3	16	15°	26	52	1.1

Parameter	l_s	l_t	w_d	l_0	w_0	s_{c1}	s_{c2}	s_{c3}	d_1	d_2
Value (mm)	13.6	4.5	0.2	17	2.3	10	7.65	6.1	4.15	3.5

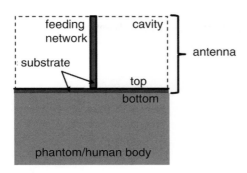

Figure 3.38 Arrangement of the antenna with differential feed network. ©2012 IEEE; reprinted with permission from [94].

The circumference of the slot determines the lowest frequency at which the surface currents concentrate around the edge of the sector-like slot, and so a resonance is excited for radiation. The parameter s_2 in Fig. 3.37(a) is important for the radiation at high frequencies since strong surface currents are observed on the two indentations there. The input impedance matching in the whole frequency band is achieved by optimizing e_1, e_2, α, s_1 and s_2. The optimized values of the parameters are given in Table 3.3. The steps between the two sectors are introduced to suppress the currents of the higher modes at high frequencies that result in significant side and grating lobes in the radiation pattern (due to the large electrical distance between the monopoles regarding the wavelength at high frequency). By using these steps, a small s_1 and s_2 can be achieved [94].

The two monopoles are fed by a differential feed network as shown in Fig. 3.37. The directions of the radiated co-polarized E-fields related to both monopoles are consistent with each other. The arrangement of the antenna and the feeding network is shown in Fig. 3.38. When attached to the human body, the feed network and the top side of the antenna are in free space while the bottom side of the antenna is directly in contact with the human body [94].

In the primary step of the optimization procedure of the antenna, the phantom ($\varepsilon_r = 20$, $\sigma = 0$ S/m) has the average relative permittivity of skin and fat at 5 GHz [52–54] and the dispersion of the phantom material is not considered. Therefore the impedance matching, and the efficiency and gain of the antenna can be investigated. Since many iterations are required, the considerations above can reduce the number of mesh cells significantly, thus reducing the computing time per iteration. In the next step, conductivity and frequency dispersion of phantom in the frequency range are included. Parameter

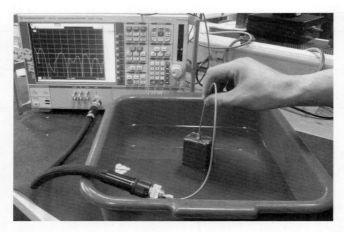

Figure 3.39 S_{11} measurement setup for the antenna prototype in the liquid medium.

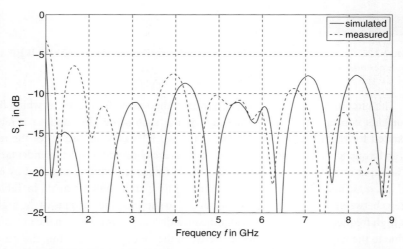

Figure 3.40 Simulated and measured S_{11} of the antenna in PEG–water solution. ©2012 IEEE; reprinted with permission from [94].

tuning is performed with the knowledge of the parameters as a fist step. Next the impedance matching and near-field pattern of the antenna are investigated; then the parameters of the antenna are updated.

3.5.3   Characterizations of UWB on-body antennas

To verify the characteristics of electromagnetic propagation experimentally inside the human body, a tissue mimicking liquid is widely used [116, 176]. As an example, in the following the characterization of the antenna from the previous section in such a tissue mimicking liquid is presented (see Fig. 3.39). Here, a mixture of polyethylene glycol (PEG) and water (weight ratio of 6 : 4) is used.

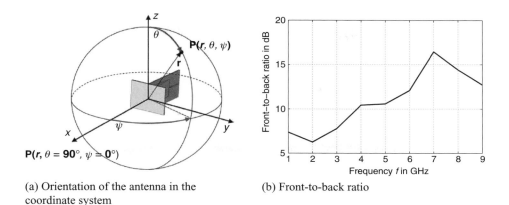

(a) Orientation of the antenna in the coordinate system

(b) Front-to-back ratio

Figure 3.41 Orientation of the antenna and front-to-back ratio of the E-field strength of the antenna at $r_0 = 30$ mm. ©2012 IEEE; reprinted with permission from [94].

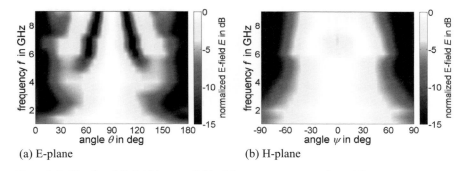

(a) E-plane

(b) H-plane

Figure 3.42 Simulated E-field in near field of the antenna over angle and frequency
(a) $E(f, \theta, \psi = 0°, r = 30\,\text{mm})$ in the E-plane for co-polarization;
(b) $E(f, \theta = 90°, \psi, r = 30\,\text{mm})$ in the H-plane for co-polarization. ©2012 IEEE; reprinted with permission from [94].

The input impedance matching of the antenna is shown in Fig. 3.40. The simulation and measurement results are very similar, in the range 1–9 GHz. The simulated S_{11} is better than -10 dB from 1.07 to 9 GHz, except for some peaks at 4.3, 7 and 8.2 GHz. It has to be noted that the feed network introduces oscillations/resonances due to the feed structure and interconnections with the antenna structure. For medical diagnostics, the antenna should radiate all its energy inside the human body, so the front-to-back ratio is a critical parameter and should be maximized. The simulated front-to-back ratio of this antenna is given in Fig. 3.41(b). The antenna is orientated as shown in Fig. 3.41(a) with its front side looking into the positive x-direction. The major reason for this large front-to-back ratio is the large ratio of the dielectric constants on the two sides of the antenna. At higher frequencies (above 4 GHz) the front-to-back ratio is even larger than 10 dB.

When evaluating antennas designed for medical applications (i.e. stroke detection) the near-field radiation must be characterized. In Fig. 3.42 the simulated E-field is shown in the near-field for the distance $r = 30$ mm. The E-field in the E-plane has a wide beam at low frequencies (up to 4 GHz) whereas the radiation pattern becomes more directive at higher frequencies. The main beam direction is quite constant at $\theta = 90°$ in the whole frequency band. Weak side lobes from 4 to 6 GHz and strong side lobes with a null at 6 GHz are also observed. A relative constant beam width is identified in the H-plane. The results show that the planar UWB slot antenna with a sector-like slot features a very stable phase center and radiation behavior in the whole frequency range from 1 to 9 GHz. By matching the antenna to the human body, an extremely small antenna and a very high front-to-back ratio can be achieved.

4 UWB antenna arrays

Grzegorz Adamiuk

A well-known method of increasing the gain and lower the half-power beamwidth of the radiation pattern is to replace the single radiator by an antenna array. Its application in communications is interesting if, for example, a point-to-point connection is to be established. It is of special interest in MIMO systems, where a channel capacity might be increased if multiple radiators, either on the transmit or the receive side (or both) are used [75]. In radar systems an application of arrays is more common in order to achieve lower half-power beamwidths, which in general are used to increase the angular resolution.

In this chapter, specific design issues of UWB antenna arrays are described. According to the methods previously described, the frequency and time domain models are explained and used for the practical array design. In the second part of the chapter an extension of the monopulse technique to UWB systems is described, based on the array theory.

4.1 Array factor in UWB systems

The resulting array radiation pattern depends on the following parameters:

- number of array elements N
- distance between the elements d
- frequency f
- excitation coefficients – amplitude and phase
- radiation pattern of a single array element $EF(f, \psi)$.

4.1.1 Array factor in the frequency domain

An antenna array usually consists of identical radiators, oriented in the same direction an equal distance d apart. In order to show the dependencies between the design parameters and radiation pattern, the arrangement shown in Fig. 4.1 can be considered. The N array elements are arranged with a spacing of d along the x-axis. The observation point P is assumed in infinite distance and varies in the $x{-}y$-plane, i.e. along the angle ψ. It is assumed that the elements do not influence the radiation properties of other elements, i.e. are ideally decoupled. In order to predict the impact of the arrangement on the radiation

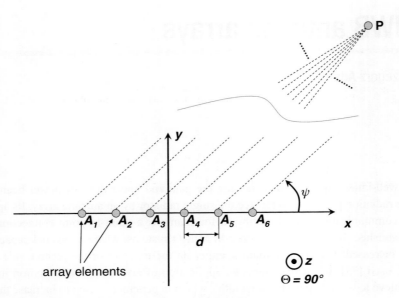

Figure 4.1 Schematic antenna array in a coordinate system.

pattern, the array factor $AF(f, \psi)$ is calculated according to

$$AF(f, \psi) = \frac{1}{N} \sum_{n=1}^{N} A_n(f) \cdot e^{-jn\varphi_0(f)} e^{j\beta \cdot n \cdot d \cdot \cos(\psi)} \qquad (4.1)$$

where n is the index of the array element, N is the number of array elements, $A_n(f)$ is the amplitude excitation coefficient of a single array element, $\varphi_0(f)$ is the frequency-dependent phase of single element, $\beta_0 = 2\pi/\lambda$ is the propagation constant in the free space, and d is the distance between the elements.

An array factor $AF(f, \psi)$ for three different frequencies for $N = 8$ elements at constant spacing $d = 5$ cm is plotted in Fig. 4.2. From the model presented in (4.1), it is clear that the array elements are assumed to be isotropic radiators. Therefore it can be observed in the diagrams that a directive radiation pattern can be achieved by the array. An important design target of the main beam is that its direction should be independent of the frequency. The beamwidth decreases as the number of array elements N, with spacing d, and/or frequency f, increases. The array factor at frequency $f = 6$ GHz and distance $d = 5$ cm is equal to the case where the same number of elements is considered at frequency $f = 3$ GHz spaced at $d = 10$ cm. Hence the array factor depends in fact on the electrical distance between the elements, i.e. the physical distance related to the wavelength.

In all cases, the side lobes are present next to the main lobe. The number and angular position of the lobes depends on the number of elements N and the frequency f. Their amplitude is related to the amplitude and phase of the excitation coefficients. For uniform excitation the amplitude of the first side lobe is approximately 21% of the main lobe,

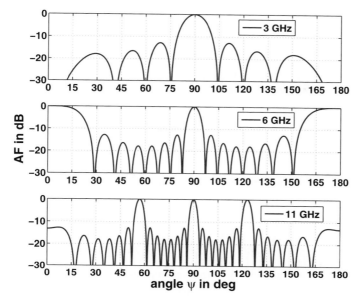

Figure 4.2 Array factor of an eight element array with 5 cm element spacing for three different frequencies.

i.e. approximately -13.2 dB. It can be lowered if, for example, a triangle or cosine taper is applied over the array aperture. For further information, see [24, 173].

Large electrical distances between the elements lead to the so-called grating lobes. Contrary to the side lobes, their amplitude is equal to that of the main lobe and their direction changes with frequency, distance between the elements, and excitation coefficients. In general, they introduce a negative effect in the radiation pattern, since unintended parts of space are illuminated with power density equal to that of the main lobe. In special applications, the grating lobes might be used for intended illumination of an additional space or signal reception from several directions simultaneously.

The resulting pattern produced by the array $C_{\text{Array}}(f, \psi)$ is the product of an array factor $AF(f, \psi)$ and the radiation pattern of the single element, called the element factor $EF(f, \psi)$ (cf. (4.2)):

$$C_{\text{Array}}(f, \psi) = AF(f, \psi) \cdot EF(f, \psi). \tag{4.2}$$

An example of a resulting radiation pattern for an array consisting of $N = 8$ elements spaced at $d = 5$ cm at frequency $f = 11$ GHz, is shown in Fig. 4.3. The top figure shows an array factor and the middle figure a directive element factor. Assuming that all elements have identical radiation patterns and the same orientation, and do not influence each other, the resulting radiation pattern is shown in the bottom figure. Note that the side and grating lobes are suppressed by the element factor. To try to keep the grating lobes away from the main lobe, the distance between the elements in the array should be kept small. This is of special importance for a UWB array, where in many cases a grating lobe-free operation over a large bandwidth is desired.

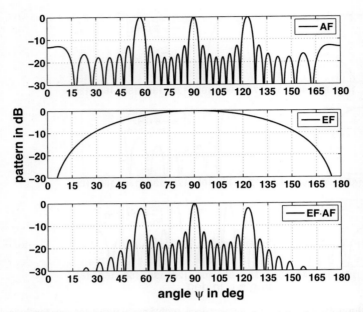

Figure 4.3 Example of resulting radiation pattern of an eight element array with 5 cm element spacing at frequency $f = 11$ GHz.

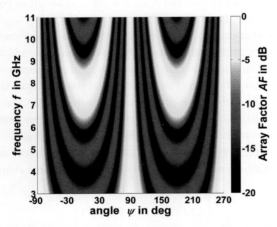

Figure 4.4 Array factor AF in the frequency domain of an array consisting of four isotropic elements spaced 4 cm apart. ©2009 IEEE; reprinted with permission from [6].

The array factor for an ultra broadband frequency range from 3 to 11 GHz for four elements spaced by $d = 4$ cm is shown in Fig. 4.4. The previously described dependencies are also valid in this case: frequency-independent main beam direction, decreasing beamwidth with increasing frequency, and side lobes with their number and position according to frequency. The grating lobes start to emerge at approximately 6 GHz and reach their full amplitude at approximately 8 GHz. This frequency corresponds to the electrical distance between the elements of exactly λ_0. Above this frequency, the grating lobes split and their position becomes closer to the main lobe with increasing frequency. For higher frequencies (not shown in the figure) or greater distances between the elements, additional pairs of grating lobes appear.

Note that the array factor is symmetrical to the x-axis (cf. Fig. 4.1), i.e. symmetrical to $\psi = 90°$ or $\psi = 270°$ in Fig. 4.4. This is an obvious consequence of the symmetrical arrangement of the array elements. In order to eliminate the second main lobe (back lobe), a single element with a directive radiation pattern must be selected.

4.1.2   Array factor in the time domain

The previously described model considers the radiation pattern only in the frequency domain. As concluded in the previous chapters, for the sufficient description of a pulse-based system, the time domain quantities are more convenient. To model an array behavior with the impulse response, an array factor $af(t, \psi)$ in the time domain is used. The $af(t, \psi)$ can be calculated by an application of the inverse Fourier transformation of the complex array factor in the frequency domain $AF(f, \psi)$. An example of $af(t, \psi)$ for the same configuration as in Fig. 4.4 (isotropic elements at distance $d = 4$ cm) is shown in Fig. 4.5. The assumption in the model is the same as before, i.e. the element factors are equal, they possess the same orientation, and do not influence each other.

The four elements interfere constructively in exactly two directions: $\psi = 90°$ and $\psi = 270°$. These directions correspond to the main beam directions in the frequency domain and the relative delay of the superposed pulses is 0 ns. The signal contributions from the single radiators are clearly resolved in directions other than the main beam directions. The number of traces is equal to the number of elements in the array. The angular delay variation of the single trace is correlated with the sine function. This is a direct result of the distance change, i.e. the path delay between the respective element and the observation point at a constant distance from the origin of the coordinate system in Fig. 4.1. The maximum delay is different for each pair of symmetrical elements, and is directly proportional to the distance between the elements and the origin of the coordinate system:

$$\max \Delta t = \frac{d \cdot \frac{n-1}{2}}{c_0} , \qquad (4.3)$$

where d is the distance between the array elements, n is the respective index of array elements, and c_0 is the speed of light. In the formula it is assumed that the array is placed symmetrically at the origin of the coordinate system.

The amplitudes of the single traces are constant over an angle ψ. The ratio between the maximum of $af(t, \psi)$ (at $\psi = 90°$ or $\psi = 270°$) to the amplitude of the single trace is equal to the total number of array elements N. It is assumed, similarly to the case of $AF(f, \psi)$, that the difference in signal attenuation (due to different distances between the array elements and the observation point P) can be neglected. The mathematical

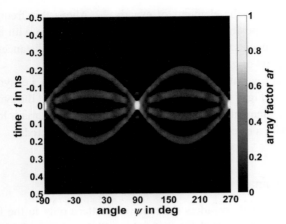

Figure 4.5 Array factor $af(t, \psi)$ in the time domain of an array consisting of four isotropic elements spaced 4 cm apart. ©2012 IEEE; reprinted with permission from [13].

model describing the array factor in the time domain $af(t, \psi)$ is as follows:

$$af(t, \psi) = \frac{1}{N} \sum_{n=1}^{N} a_n(t) \cdot \delta(t + \tau_d(\psi) + n \cdot \tau_e) \tag{4.4}$$

where δ represents a Dirac pulse, $a_n(t)$ an amplitude excitation coefficient, $\tau_d(\psi) = \max \Delta t \cdot \cos(\psi)$ is the delay of the radiated pulse in the specified direction ψ depending on the position of an array element, and τ_e is an intentionally applied delay difference between the array elements for the purpose of beam shifting (in Fig. 4.5, $\tau_e = 0$ ps). The case of beam shifting ($\tau_e \neq 0$ ps) is discussed in the following sections.

The resulting radiation pattern of an array in the time domain is a convolution of the impulse response of the single array element $h(t, \psi)$ with the array factor $af(t, \psi)$. Similarly to the array factor in the frequency domain $AF(f, \psi)$, the back lobe can be suppressed by a directive characteristic of the single radiator.

In the frequency domain, grating lobes arise in the array factor $AF(f, \psi)$ at increasing electrical distance between the array elements. The grating lobes possess the same amplitude as the main beam and their direction is frequency-dependent. If the array factor in the time domain $af(t, \psi)$ is considered, grating lobes do not appear to be created. In the direction other than the main beam, the pulses resulting from single array elements are spread over the time. The maximum pulse spread occurs for the directions $0°$ or $180°$ (cf. (4.3)), i.e. in the direction of the array extension. Similarly to grating lobes, the amplitude of the spread pulses can be suppressed by the radiation pattern (element factor) of the array elements.

4.1.3   Shift of the beam

By applying a constant offset in the phase excitation between neighboring array elements, the main beam direction of an array can be shifted at just one single frequency. In the

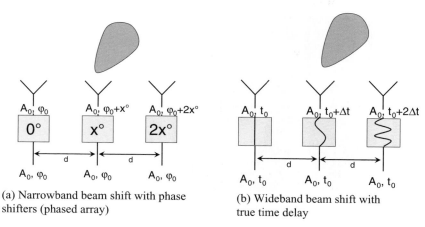

(a) Narrowband beam shift with phase
shifters (phased array)

(b) Wideband beam shift with
true time delay

Figure 4.6 Principles of beam shifting.

case of narrowband systems, the phase shift φ_0 might be realized by simple phase shifters,
which are designed to provide a constant phase shift, independent of the frequency
[24, 173] (see Fig. 4.6(a)). In a UWB array this method could lead to a frequency
dependency in the beam shift. In order to overcome this problem a true time delay
(TTD) beamforming network is applied (cf. Fig. 4.6(b)). TTD delivers, in contradiction
to the standard phase shifters, a constant delay offset τ_e of the signal between the
neighboring array elements (cf. (4.4)). This leads to a linearly increasing phase shift
$\varphi_0(f)$ over the frequency for a single element (cf. (4.1)):

$$\varphi_0(f) = \tau_e \cdot f \cdot 360°, \tag{4.5}$$

where τ_e is the signal delay offset between the neighboring elements, f is frequency and
$\varphi_0(f)$ is expressed in degrees.

The relationship between the direction of the main beam ψ_{mb} and the signal delay
offset τ_e is

$$\psi_{mb} = \arccos\left(\frac{c_0 \cdot \tau_e}{d}\right). \tag{4.6}$$

Note that the main beam direction depends solely on the delay offset τ_e in relation to the
distance between the elements d.

There are several possibilities for the practical realization of true time delay beamforming networks. The simplest solution is an application of switched microstrip lines
with different lengths for each array element [25]. A more sophisticated solution are
Finite Impulse Response (FIR) filters [114]. FIR filters offer the additional possibility of
flexible adjustments to the excitation signal amplitude, but suffer from high complexity.
Another method of TTD beamforming is the application of Rotman lenses [85]. This
method offers rather low flexibility with the design and, in most cases, suffers from low
efficiency. However, depending on the technology in which the lens is implemented,

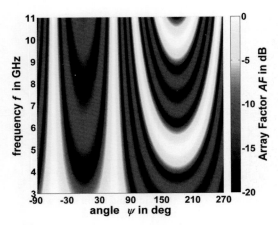

Figure 4.7 Array factor $AF(f, \psi)$ in the frequency domain of an array consisting of four isotropic elements spaced 4 cm apart; delay between the elements $\tau_e = 70$ ps.

it might offer a high power capability – which is especially interesting for military or security applications.

Frequency domain

The array factor in the frequency domain $AF(f, \psi)$ for a TTD shifted beam is shown in Fig. 4.7. The array consists of four elements with spacing of 4 cm. The implemented delay offset between the elements is $\tau_e = 70$ ps. The main beam has been shifted from the direction of 90° to approximately 58°. The back lobe has been shifted from −90° to approximately −58°, while the direction of the main and back lobes is constant over the frequency. This implies minimum pulse distortion caused by the array in this particular direction. Note that the side and grating lobes are, in the case of beam scanning, unsymmetrical w.r.t. the main beams. For larger scan angles, the grating lobes may move towards the original direction of the main beam (i.e. before scan). This might be critical for particular applications, since the grating lobes are no longer suppressed by the radiation pattern of the single array element.

Time domain

The corresponding array factor in the time domain $af(t, \psi)$ is presented in Fig. 4.8. The radiation is coherent for only two directions ±58°, which conform with the main beam directions of $AF(f, \psi)$ in Fig. 4.7. For all other directions, the signals originating from the single array elements are resolvable. Similarly to the grating lobes, these traces are unsymmetrical w.r.t. the directions of coherent radiation. The local maximum signal spread over time is for the angles 0° and 180°, similar to the no-scan case in Fig. 4.5. However in the case of scanning, the local maxima of a pulse spread have different values. The shortening or lengthening of the maximum spread is proportional to the implemented delay offset τ_e.

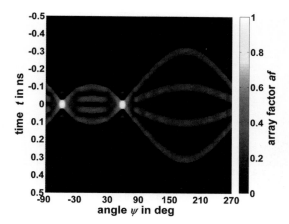

Figure 4.8 Array factor $af(t, \psi)$ in the time domain of an array consisting of four isotropic elements spaced 4 cm apart; delay between the elements $\tau_e = 70$ ps.

4.1.4 Radiation characteristics of a real UWB array

In this section, examples of the resulting radiation patterns in the frequency and time domains are given for an array. It is assumed that the array consists of $N = 4$ directive, dual-polarized differentially fed 4-ellipse antenna elements – see the photograph in Fig. 3.29(a) and radiation pattern in the frequency domain (Fig. 3.30(a)) and in the time domain (Fig. 3.30(b)). The main beam direction of the single antenna is in the direction of the positive y-axis, and the elements are distributed along the x-axis as shown in Fig. 4.1. The distance between the elements is $d = 4$ cm, which is the minimum possible value due to the transversal dimension of the single antenna element. The array is fed by a 4-way (6 dB) power divider, which is connected with the radiators via cables. The losses in the feeding network are taken into account by means of the measured transfer function, which includes all losses and signal distortions.

Frequency domain

The model also includes all losses of the radiator, which are included in the measured radiation pattern (in both time and frequency domains). The elements are assumed as ideally decoupled and identical. With these assumptions and boundary conditions, the resulting complex transfer function of the array $H_{ar}(f, \psi)$ is calculated as:

$$H_{ar}(f, \psi) = H_{ant}(f, \psi) \cdot AF(f, \psi) \cdot H_{feed}(f) \tag{4.7}$$

where $H_{ant}(f, \psi)$ is the complex transfer function of the single array element, $AF(f, \psi)$ is the complex array factor, and $H_{feed}(f)$ is the complex transfer function of the feeding network.

The resulting gain of the array $G_{ar}(f, \psi)$, which is calculated from $H_{ar}(f, \psi)$, is shown in Fig. 4.9(a) for $\tau_e = 0$ ps and in Fig. 4.9(b) for $\tau_e = 70$ ps. In the case of no scan, the back lobe is significantly suppressed by the directive radiation pattern of the array elements. The amplitudes of the grating lobes are also suppressed. They make

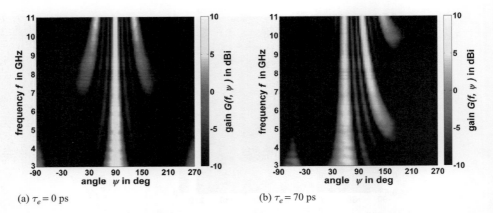

(a) $\tau_e = 0$ ps   (b) $\tau_e = 70$ ps

Figure 4.9 Gain $G_{ar}(f, \psi)$ in the frequency domain of an array consisting of four elements spaced 4 cm apart. The complete antenna from Fig. 3.29(a) is used as an element. (a) ©2009 IEEE; reprinted with permission from [6]; (b) ©2012 IEEE; reprinted with permission from [13].

a significant contribution to the gain pattern above approximately 7 GHz, where the radiation pattern of the single element is not directive enough to provide a suppression in the relevant angular region. Note that the maximum gain is approximately 10 dBi. According to the theory, its increase should be exactly 6 dB for $N = 4$. This would result in the maximum gain of the array in the region of 11 dBi, which is then reduced by losses in the feeding network which are included in the model. It can be concluded that by the application of an array, the beamwidth can be significantly reduced compared to the pattern of the single elements in Fig. 3.30(a).

In the case of beam scanning, the main lobe is clearly present at approximately 58°. Its main beam direction is constant, as expected from the theory. Its amplitude is slightly reduced in comparison to the case of no scan, which is a result of the influence of the element factor (in (4.7), included in $H_{ant}(f, \psi)$). The back lobe direction is, in this case, also significantly suppressed, however the grating lobes are fully within the illuminated area of the single radiator and hence cannot be avoided. This introduces a significant limitation if the array is to be used for narrowband signals distributed over a large bandwidth.

Time domain

The resulting impulse response of the array $h_{ar}(t, \psi)$ is a convolution of the impulse response of the respective array components and is calculated as follows:

$$h_{ar}(t, \psi) = h_{ant}(t, \psi) * af(t, \psi) * h_{feed}(t). \tag{4.8}$$

$h_{ant}(t, \psi)$ is the impulse response of the single array element, $af(t, \psi)$ is the array factor in the time domain and $h_{feed}(t)$ is the impulse response of the feeding network. The impulse responses for the same scanning conditions are shown in Figs. 4.10(a) and 4.10(b), respectively. The impulse response is significantly delayed in both cases, and the maximum is approximately 3.5 ns. The major contributions to the cumulative

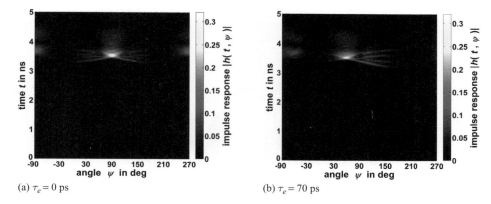

(a) $\tau_e = 0$ ps   (b) $\tau_e = 70$ ps

Figure 4.10 Impulse response $|h(t, \psi)|$ in the time domain of an array consisting of four elements spaced 4 cm apart. The complete antenna from Fig. 3.29(a) is used as an element. (a) ©2009 IEEE; reprinted with permission from [6]; (b) ©2012 IEEE; reprinted with permission from [13].

delay are: (i) the delay of the single elements (cf. Fig. 3.30(b)); and (ii) the delay caused by the feeding network (power divider and 50 cm long cables). There is exactly one direction of coherent radiation, which is compliant with the position of the main beam in the frequency domain characteristics. The second direction of the coherent radiation, which is observed in $af(t, \psi)$, is suppressed by the impulse response of the array element. Apart from these directions, the contributions from single array elements are resolved. According to theoretical calculations, the maximum spread of the pulse is reduced by the radiation characteristic of the single radiator.

Conclusion

In the design of a UWB antenna array, the spacing between array elements should be kept as small as possible. In many cases the minimum distance is determined by the maximum dimension of the single radiator. This often has to be $0.5\lambda_0$ at the lowest frequency in order to satisfy the radiation conditions. Such a distance leads to grating lobes at higher frequencies, which are hard to suppress by the element factor. At this point we need to distinguish between the usage of narrowband signals (which are switched over large bandwidths) and pulse operation. In the case of narrowband signals, the generation of grating lobes at high central frequencies might be crucial. In the pulse operation (i.e. instantaneous broadband), grating lobes do not get generated at all. The large distance between the elements merely causes a larger spread of the pulse outside of the main beam, which is less critical than concentrated grating lobes.

4.2   UWB amplitude monopulse arrays

Monopulse antenna arrays are able to work in at least two radiation modes. The aim of implementation of the different modes is to create a dependency between the transmitted (or received) power and the angle of radiation (or reception). A look-up table is

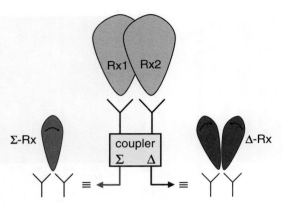

Figure 4.11 Principle of amplitude monopulse for the receive case: creation of sum and difference beams. The arrows denote if the beam's phase is shifted by 180°.

generated by measuring the radiation pattern of the antenna. The look-up table delivers an unambiguous dependency between the power ratios in different modi and the angle for a particular angular region. By comparing the processed target echoes received in multi-mode operation with the look-up table, a precise angular position of the target can be determined [156]. The information for the determination of the target direction might be contained in the phase (phase monopulse technique), the amplitude (amplitude monopulse technique), or both (hybrid monopulse technique). To determine the target angular position in a two-dimensional plane, one sum (Σ) and one differential (Δ) mode are usually sufficient. The heart of such a system is the feeding network/coupler, which separates the sum and differential signals, i.e. it produces a sum and a differential beam. The monopulse technique requires a priori no bandwidth and can generally be applied in monofrequency systems. However in order to achieve a higher range resolution, the bandwidth of the signal must be increased. A combination of the monopulse technique and UWB delivers the unique possibility of simple direction estimation of the radar target with simultaneous superior range resolution. This performance is achieved with a single radar echo and very little processing. In low resolution radars, simple 180° hybrid couplers, like rat-race couplers, are sufficient as feeding networks. For UWB, some special monopulse frequency-independent feeding networks and broadband antennas have to be combined.

As mentioned before, the core component of a monopulse system is the feeding network. In this case a UWB 180° hybrid coupler has to be integrated. Such couplers possess two input and two output ports. In the receive case, the two input ports are connected to the respective radiators. The two output ports are named the sum (Σ) and the differential (Δ) ports. In the ideal case, both ports are completely decoupled from each other. The key functionality of the coupler is the addition of the in-phase and the out-of-phase signals received through the radiators as shown in Fig. 4.11. In the case of signal reception at the Σ-port, both input signals are summed up coherently. This results in a Σ-beam, which represents the most typical case, as described in the previous section. During signal reception at the Δ-port, the signals are summed up with a phase

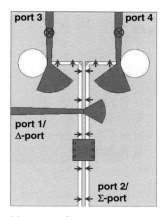
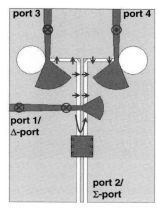

(a) sum mode
(in-phase output signals)

(b) differential mode
(out-of-phase output signals)

Figure 4.12 Principle of $\Sigma\,\Delta$-hybrid coupler. ©2010 KIT Scientific Publishing; reprinted with permission from [3].

difference of 180° between the input signals coming from the antennas. In UWB systems the in-phase and out-of-phase signal combination must be performed in a frequency-independent way, i.e. the respective phase shifts should be valid for each frequency within the dedicated bandwidth [10].

In the Σ-mode, a single beam is created in the main direction, which is narrower than the beam of a single antenna. However in the Δ-mode, a radiation pattern with two beams positioned symmetrically on both sides of the main axis is generated. The angular position of the notch between the beams in Δ-mode is exactly at the maximum of the Σ-beam. Due to the physical properties of radiation, the forms of the beams and their widths change over the applied frequency range. This can be tolerated for UWB monopulse systems, as long as the maximum of the Σ-beam and the notch of the Δ-beam remain constant in the angular direction. This is always fulfilled if the previously mentioned requirements regarding the phase differences between the input signals are met. Some practical design results of a UWB monopulse array will now be discussed.

UWB 180° hybrid coupler

An example of a frequency-independent coupler for UWB monopulse applications is shown in Fig. 4.12. The dark areas indicate the metalization on the top of the coupler and the light at the bottom. Two input ports are indicated in the figure as port 1 (Δ-port) and port 2 (Σ-port). An input signal is transfered from one of the feeding ports to the output ports 3 and 4. The signals at the outputs have the same amplitude and are in-phase when port 2 is excited. A schematic electric field distribution in the coupler structure in the excitation case at the Σ-port is shown in Fig. 4.12(a). The signal is guided in two adjacent slots in the coplanar waveguide (CPW) mode and is transfered farther through the set of vias. The vias take a shortcut through the outer metalizations, which is realized by the metalized patch placed on the opposite side of the substrate. Since the outer

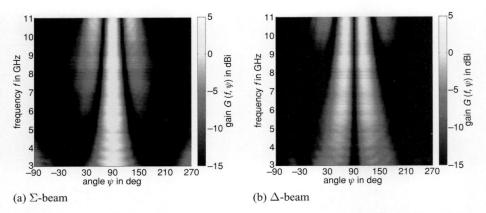

(a) Σ-beam   (b) Δ-beam

Figure 4.13 Measured gain in the frequency domain of a monopulse UWB array consisting of two elements spaced 4 cm apart. The complete antenna from Fig. 3.29(a) is used as an element. ©2010 IEEE; reprinted with permission from [5].

metalization possesses the same potential, the vias do not have a significant influence on the propagation of the CPW mode. In the next step, adjacent slots are separated into simultaneous slotlines. Each of the slotlines is then transformed by an aperture coupling to the microstrip line. From the orientation of the electric field lines, it is possible to conclude that the signals at ports 3 and 4 possess the same phase.

The signals at 3 and 4 are differential if port 1 (Δ-port) is excited. The electric field is guided through the microstrip line to the center of the structure (see Fig. 4.12(b)). Next, the energy is coupled to the two adjacent slots by an aperture coupling and the couple slotline (CSL) mode is excited. It results in the excitation of both slots with the electric fields oriented in the same direction. The coupled electric fields are propagated in the direction of the outputs. The directional propagation of the fields to the outputs is achieved by an implementation of the vias. The vias at the outer metallization with the connecting patch introduce a shortcut for the CSL mode and reflect it.

Monopulse beams in the frequency domain

The principle of creation of Σ- and Δ-beams is explained based on an array with $N = 2$ elements at a distance of $d = 4$ cm from each other. The two elements are placed on the x-axis and symmetrically to the y-axis, according to Fig. 4.1. Also, in this case, the complete antenna from Fig. 3.29 is used as an array element. Since the array extension is only performed in one plane, the monopulse technique can only be applied in this case for a two-dimensional scenario. If additional array elements in a proper orientation are applied, an extension of the monopulse technique for the exact localization of the target in all three dimensions is possible. This approach is identical to the case of narrowband systems, and the reader is therefore referred to the other literature, e.g. [156].

The measured frequency domain patterns of the two-element UWB monopulse array in the Σ- and Δ-modes are shown in Figs. 4.13(a) and 4.13(b), respectively [5]. It can be seen that the shapes of the beams change their form over the frequency. However the Σ-mode has a constant maximum (main beam direction) of 90° and, in the same

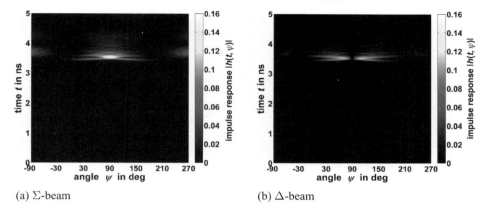

(a) Σ-beam   (b) Δ-beam

Figure 4.14 Measured impulse response of a monopulse UWB array consisting of two elements spaced 4 cm apart. The complete antenna from Fig. 3.29(a) is used as an element. ©2010 IEEE; reprinted with permission from [5].

direction, the Δ-mode has a strong notch in the radiation pattern. Also, the position of the notch is nearly constant over the frequency. Such a set of beams can be successfully applied for the monopulse radar over a large frequency range.

Monopulse beams in the time domain

In IR-UWB, the information for a single frequency cannot be extracted from the received signal, though one single value characterizing the received UWB signal (i.e. also the radiation pattern for the whole bandwidth) is needed in order to perform the monopulse direction finding. For this purpose, a maximum of the impulse response in both modes for a particular direction can be used [8]. The measured impulse response of the presented example is shown in Figs. 4.14(a) and 4.14(b) for Σ- and Δ-modes, respectively. Likewise, as in Fig. 4.13, the maximum for the Σ-mode is in the same direction as the notch of the Δ-mode, i.e. at $90°$. The relatively large delay of the pulse is caused by the connecting cables between coupler and antennas, and also by the coupler and antennas themselves.

Look-up table

In the following, an approach for the determination of the look-up table is described using the ratios between the radiation patterns in the Σ- and Δ-modes. The ratio of the amplitudes of the target echoes received by the radar in both modes are exactly the same as in the look-up table for the specified directions. This makes it possible to obtain an exact estimation of the target direction. An evaluation solely of the amplitude of the radiation patterns leads to an ambiguity in the angular range. The ratio of the amplitude is exactly the same for the angles symmetrical to the main beam direction. This allows a precise position estimation, without differentiation between directions symmetrically left and right of the main beam direction. In order to get rid of this ambiguity, the phase information has to be evaluated. The phase of the signal in the Σ-beam remains constant for the whole beamwidth. In the Δ-mode the phase of the signal contained in both beams

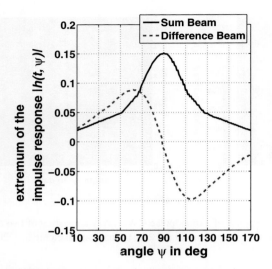

Figure 4.15 Extrema of the measured impulse responses $|h(t, \psi)|$ of an array consisting of four elements spaced 4 cm apart. The complete antenna from Fig. 3.29(a) is used as an element.

is exactly opposite, i.e. $180°$. Summarizing, the maximum of the impulse response $\max(|h(t, \psi)|)$ is used for finding the direction, and the phase information allows for the left–right recognition of the target direction. The phase information contained in the signal may be extracted by, for example, the leading sign of the correlation coefficient. As a reference pulse, the real part of the impulse response in the main beam direction in the Σ-beam $h_\Sigma(t, \psi = 90°)$ can be used. The mathematical description for the extraction of the phase information $\xi(\psi)$ from the impulse responses yields:

$$\xi(\psi) = \text{sgn}\left(\text{corr}\left(h_\Sigma(t, \psi = 90°), h_\Delta(t, \psi)\right)\right), \tag{4.9}$$

where $h_\Sigma(t, \psi = 90°)$ is the impulse response of the monopulse array in Σ-mode in the main beam direction $\psi = 90°$ and $h_\Delta(t, \psi)$ is the impulse response of the monopulse array in Δ-mode. For an ideal monopulse array, the values of $\xi(\psi)$ are

$$\xi(\psi) = \begin{cases} -1, & \text{for } \psi < 90° \\ 1, & \text{for } \psi > 90°. \end{cases} \tag{4.10}$$

Using the phase information, the extrema of the impulse responses in both modi of the monopulse array can be plotted. These are mathematically described by:

$$p_\Sigma(\psi) = \max_t |h_\Sigma(t, \psi)| \tag{4.11}$$

$$p_\Delta(\psi) = \xi(\psi) \cdot \max_t |h_\Delta(t, \psi)|. \tag{4.12}$$

The extrema of the impulse responses $p_\Sigma(\psi)$ and $p_\Delta(\psi)$ containing phase information are plotted in Fig. 4.15 over the angle ψ. In the Σ-mode case, a similarity between the radiation pattern in the frequency and in the time domain is observed – with a single

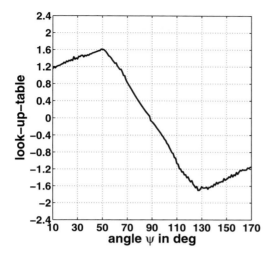

Figure 4.16 Look-up table of the monopulse array consisting of four elements spaced 4 cm apart. The complete antenna from Fig. 3.29(a) is used as an element.

beam and beam direction at 90°. In the case of the Δ-beam, $p_\Delta(\psi)$ takes both positive and negative values. The ratio between $p_\Delta(\psi)$ and $p_\Sigma(\psi)$ creates the look-up table $\Xi(\psi)$:

$$\Xi(\psi) \equiv \frac{p_\Delta(\psi)}{p_\Sigma(\psi)} . \qquad (4.13)$$

The look-up table, calculated as above, is plotted in Fig. 4.16. It can be clearly seen that there is an unambiguous range around the angle $\psi = 90° \pm 20°$. For larger angles, $\Xi(\psi)$ becomes ambiguous and it has to be guaranteed by other means (analog, or by additional algorithms) so that the target echoes coming from those angles are distinguished from the ones received in the unambiguous range. Comparing the calculated value (based on the target echoes) with the look-up table in Fig. 4.16, the direction of the target can be found. The technique is only valid for one single radar echo within the respective portion of time. Different time delays of echoes coming from different targets are resolved by the UWB technique in time (and thereby in range as well), and do not generally have an influence on detection performance. If echoes from two or more targets are received at the same time, a larger number of modes in the monopulse array or beam scanning can be used in order to distinguish the targets from each other. The accuracy of the direction estimation depends mainly on the receiver sensitivity and (mechanical and electrical) array design.

5 Monolithic integrated circuits for UWB transceivers

Gunter Fischer and Christoph Scheytt

The great interest in UWB systems is well-motivated by the huge signal bandwidth. As shown in the previous chapters, the ultra-wide signal bandwidth has many benefits, such as very high data transmission rates and very fine time resolution. Either or both of these potentials would be very attractive for short-range communication systems, if a substantial advantage in terms of power consumption or performance could be achieved compared to existing radio solutions. In particular, the possible time resolution well below 1 ns is a unique feature, which seems to be the enabler for precise indoor localization. Unfortunately, the huge signal bandwidth demands an appropriate circuit design, which in many cases differs from traditional narrowband RF circuit design. The same requirement holds for the sub-circuits in the digital domain in order to obtain the desired timing accuracy. Having indoor localization applications in mind, this chapter shows why the impulse radio UWB circuits are promising in terms of localization accuracy in time-of-arrival (TOA)- and time-of-flight (TOF)-based localization systems. Different pulse generation and pulse detection principles will be introduced and explained. Important design considerations will be discussed on the way to a fully monolithic integrated circuit. Finally, there are a few examples for illustration.

5.1 Pulse radio transceiver requirements

5.1.1 Indoor channel requirements

Why did recent short-range radio systems like Bluetooth, WLAN, ZigBee, not accomplish a breakthrough in indoor localization? Summarizing the indoor channel properties from Chapter 2, the answer lies in their limited signal bandwidth which does not allow them to resolve the multipath propagation dominating in indoor environments. Figure 5.1 illustrates again a typical situation in a room. A transmitter emits radio waves, which either travel directly to a receiver (line-of-sight (LOS) situation) or will be reflected somewhere and appear later at the receiver (non-line-of-sight (NLOS) situation). A typical radio communication system can more or less deal with that, since the exact time of arrival of the signal does not matter and potential fading effects are compensated for by more receiver gain. With localization systems the situation is more severe, since additional delays in the reception of signals traveling on paths apart from the LOS path may lead to a wrong time of arrival estimation – and therefore result

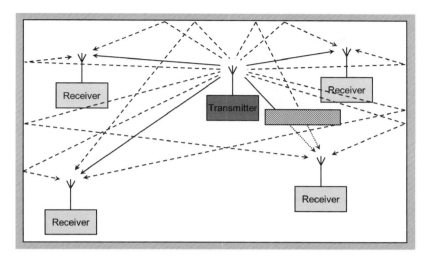

Figure 5.1 Illustration of LOS paths (solid line) and NLOS paths (dashed line) in indoor localization applications.

in a failure in distance calculation. In order to distinguish between the most relevant LOS path and the other NLOS paths, one needs somehow to resolve the superposition of received signals. This is difficult for narrowband systems, where the duration of the transmitted symbol is usually much longer than the typical excess delays of the different propagation paths. The strongest path (due to the mentioned fading effects, but not necessarily the LOS path) will finally dominate the receiver output. At this point, the impulse radio transmission benefits from the fact that pulses are typically much shorter than excess delays of NLOS paths. In the time domain the received pulses can be separated and assigned to different propagation paths. As already emphasized, the identification of the first incoming signal (pulse), which usually travels along the LOS path, is of particular importance for the accuracy of the distance calculation.

5.1.2   Timing accuracy for indoor localization

A broad overview of UWB localization techniques will be given in Section 6.2. Here, only time-of-flight based methods are considered due to their significant influence on transceiver architectures. These kinds of localization techniques basically require the capability to determine the time-of-transmission (TOT) of an emitted radio signal and the time-of-arrival (TOA) of its reception in order to determine the propagation time. Of course, the accuracy of the calculated distance between two radio nodes is strongly affected by the raw time measurement resolution. Remember, since radio waves propagate at the speed of light, an uncertainty in time measurement of 1 ns would correspond to an uncertainty of 30 cm in distance determination. This may already matter in most applications, since several single errors may contribute independently to the overall localization variance corresponding to the applied method. For instance, taking the two-way ranging (TWR) scheme, the uncertainty of each of the two involved nodes will be added.

5.1.3   Example consideration

In order to roughly quantify the timing requirements of an indoor localization system, let's assume a generic application as follows. People (moving with a walking speed of 1 m/s) shall be guided by a navigation system in a large building (e.g. fair, airport). Each of 100 users may demand an update of their individual position once per second with an accuracy of 1 m. Further assume that all users share the same RF medium (channel) which probably leads to a time division multiple access (TDMA) scheme. Here only 10 ms per user are available for the entire localization procedure. Depending on the chosen localization scheme, this procedure consists of several independent transmissions between mobile and anchor nodes.

To simplify matters, without going into detail, one can estimate that one transmission of a localization data packet should not take longer than 1 ms. From the laws of statistics, we know that increasing the number of samples would decrease the variance by a square law, as long as the boundary distribution is reached. For a given time measurement interval, this means that any growth of the sampling rate and any widening of the signal bandwidth will improve the accuracy. In relation to our example, one can state that 1000 samples (e.g. zero crossings) of a received signal sampled with a rate of 10 Msamples/s would be sufficient to achieve a quantization error of less than 3.3 ns. This only holds if the received signal is periodic but not correlated to the sampling system *and* it is not disturbed by any noise.

However, in a real life system several significant error sources will contribute to the overall error. First of all the noisy transmitter with the timing based on a noisy local oscillator will contribute an initial error distribution. Additional noise from the transmission channel as well as from the receiver circuitry will broaden the error distribution of the signal which finally appears at the sampling system (e.g. AD converter). Here, the jitter of the local clock will again increase the error distribution function. Since all the error contributions add up, none of them should exceed the required overall accuracy. For instance, a poor transmit signal cannot be improved, not even by an ideal transmission and sampling system. Also, the inherent clock jitter of the sampling system cannot be compensated afterwards. In the case of poor SNR of the transmission channel, one may improve the statistics by increasing the number of samples (by widening the signal bandwidth or prolonging the time frame of the measurement) up to an allowable amount, which leads to a reduced error contribution of uncorrelated noise.

5.1.4   Implications for circuit design

From the considerations above one can conclude that the availability of ultra-wide signal bandwidth of UWB systems allows a tremendous reduction of the localization error induced by the wireless transmission and general uncorrelated noise. Assuming sufficiently accurate timing at the transmitter and receiver, bandwidth is the limiting factor for today's radio systems. The emerging UWB systems allow not only much better localization resolution but also a much larger diversity of applications – like higher update rates per mobile node, or increased number of nodes operating simultaneously in the system. For instance, localization systems involving more than 1000 nodes or update rates of more than 1000 Hz become feasible. In addition, the use of short pulses (less than 1 ns) as transmission medium (carrier) substantially simplifies coping with multipath propagation in indoor environments. This makes pulse radio UWB systems superior for indoor localization applications. Whereas other large-signal bandwidth systems like OFDM achieve similar data rates in data communication, they hardly compete with pulse radio systems in terms of indoor localization. The ultimate solution for many unsatisfied demands could be a combination of high data rate communication with precise indoor localization using only one radio frontend – while consuming less power than two single solutions. The pulse radio UWB circuits could be the perfect answer if a number of RF circuit design challenges can be met. For example, the huge signal bandwidth requires different techniques in impedance matching and it makes a receiver much more susceptible to noise and other interfering radio signals. Furthermore, the tight timing requirements for high-precision localization necessitate an on-chip synchronized operation of high-speed digital logic with RF circuitry while minimizing mutual disturbances. These particular aspects of RF circuit design for pulse radio transceivers intended for indoor localization applications will now be discussed.

5.2   Pulse generation

5.2.1   Generation concepts

In order to generate pulses with an ultra-wide bandwidth of 500 MHz or more, their duration has to be limited to around 2 ns or even less. In earlier times this was hard to achieve. At the beginning of the last century spark-gap transmitters were used for wireless telegraphy. They were later prohibited due to their excessive disturbance of other radio systems. Due to the progress of semiconductor technology, new devices – the so-called "step recovery diodes" (SRD) – were developed, which could generate short pulses. Unfortunately, important parameters such as pulse duration and spectral power distribution are set by the fabrication process and cannot be adapted during operation. This became an important drawback after recent regulations regarding the usage of UWB techniques (see Chapter 1). The regulations restrict the usage to certain frequency bands and limit the emitted RF power within these bands to maximum values. The typical behavior of SRDs is not favorable in this context because most of the generated RF power lies outside the UWB band. One could apply band pass filters to fulfill the regulatory requirements but the spectral efficiency and the overall power dissipation of such a pulse generator would be degraded. A more modern attempt avoiding SRDs can be found in [95].

Recent developments take advantage of the progress in silicon technologies. In order to improve the power efficiency of pulse generation it is desirable to generate the RF

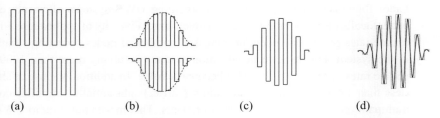

Figure 5.2 Main steps of all-digital pulse generation: (a) clock generation; (b) pulse shaping; (c) "DC-free" combination of the half monocycles; (d) spectrum smoothing by filtering.

power directly in the band of interest. Furthermore, dedicated pulse shaping (in the time domain) needs to be implemented to prevent any violation of the spectral masks given in the regulations (see Chapter 1). For the selection of an appropriate silicon technology for monolithic integration, the cut-off frequency of active devices (transistors) has to be sufficiently high and the expected passive components can be integrated. Finally, the peak voltage swing required at the antenna input to allow the expected emission of pulses with a certain power may also influence the choice of silicon technology. Taking this into account, two main approaches for pulse generation can be found in recent literature [189]. First, direct pulse synthesizing in fast digital logic; second, the up-conversion of pulse with well-defined shape into the RF domain. Both approaches, sometimes called carrierless and carrier-based, respectively, have their advantages and drawbacks and will be further investigated here.

5.2.2   All-digital pulse synthesis

Let's start with the approach mentioned first, the all-digital solution. The main idea is that the whole pulse waveform will be constructed from a sequence of successive half monocycles modulated in amplitude, where the cycle duration corresponds to the expected center frequency of the finally emitted pulses. Figure 5.2 illustrates the main steps of the entire procedure. The advantage of such strategy of pulse generation is the avoidance of passive RF components like coils or varactors (in voltage-controlled oscillators (VCOs)) as required in the second principle. Recent examples in literature can be found in [83, 103, 117].

One can imagine that the generation of the main RF signal in the digital domain is not that simple due to the high frequency in the range 3.1–10.6 GHz. If the target silicon technology allows digital clocks in that frequency range, a typical clock generator can deliver the main clock ideally in two opposite phases. The requirements on the accuracy of this clock are set by the application. For instance, a precise localization system may require some maximum levels for clock jitter and the frequency stability affects the spectral mask of the transmitter signal. Since usual process, supply voltage and temperature (PVT) variations heavily influence the output frequency of a free-running digitally controlled oscillator (DCO), a feedback loop is required to stabilize the clock.

Clock dividers (counters) are then needed within the loop, which typically consume a significant amount of power.

Another implementation approach embeds ring oscillators or delay cell rings in a delay-locked loop (DLL), where the single delay time corresponds to the half-cycle time of the final pulse. Again, the DLL is required due to the sensitivity of delay elements to PVT variations, even if a variation of the pulse center frequency could somehow be tolerated in terms of the spectral power distribution (violation of the UWB mask). Preferably, the oscillator is also used as a system clock to provide further timing events (time frames, pulse release time, etc.), where any gross variation may lead to difficulties in synchronization and communication procedures.

The pulse shaping is executed in two general tasks: first, the definition of spectral distribution of the RF power in the frequency domain (in accordance with UWB regulations and possible compliance to a standard) and second, in many cases the determination of the time of transmission of a single pulse (time of release). This can be done by applying (multiplying) the desired envelope of the pulse to the clock signal. As depicted in Fig. 5.2(b), the envelope cuts out a small fraction of the clock signal which is further processed. In fact, this ends up in an amplitude modulation of the raw clock half cycles, which will be later merged into the final pulse shape. While the general pulse shape (envelope) will not change, it lends itself to be implemented into the RF frontend in a fixed manner. In the literature, solutions with fast digital–analog converters (DACs) can be found, which are fed by a built-in look-up table or by a fixed logic combination creating a sequence of proper amplitude levels [83, 117]. It is clear that the discrete levels should be as precise as possible to reduce the unwanted spikes in the frequency spectrum possibly violating the spectral mask. As a trade-off, providing only a few amplitude level steps leads to more effort in filtering in the RF domain at the transmitter output, which may have a negative impact on the overall power dissipation. In particular, the transmitter output is usually matched to 50 Ω impedance, which demands much larger DC currents than commonly available in the digital domain.

The positive and negative half cycles constructing the final pulse are typically generated in two separate branches for simplicity. However, the RF pulse has to be DC-free. Therefore, a proper combination of the two separate signals is needed, which removes the DC content of the signal. The simplest way to do this is to use an AC coupling via capacitors. Note that these capacitors act as a high pass filter and need to be considered in terms of pulse shaping. Similar considerations are needed for types of DC offset compensations to remove the DC part. At this point a binary phase-shift keying modulation can be easily incorporated by simply swapping these two signals in the following stage of the transmitter chain. In many cases the two separate (and still digital) signals have rail-to-rail voltage levels provided by CMOS logic outputs. It would be wise to conserve these levels in order to realize maximum power efficiency. Therefore, the combination of the two signals is often done within the power amplifier (PA) itself, where one branch delivers positive half cycles and a second delivers negative half cycles to the common load. This allows peak voltage levels up to twice the CMOS supply voltage (the previously mentioned rail-to-rail voltage). The use of numerous CMOS inverters as PA output stages is typical for all-digital solutions in order to achieve the necessary driver capability to the RF load (i.e. antenna). Furthermore, the driver capability might be controlled by the desired pulse envelope to introduce amplitude modulation.

The following load consists of a matching network connecting the transmit antenna to the PA output. Special attention has to be paid to the fact that the matching network also acts as band pass filter, finally assuring the spectral purity of the transmit signal. The nature of the all-digital approaches introduces many undesired spikes and ripples within the frequency spectrum caused by clocks, quantization effects, switching actions, and so on. Whereas the out-of-band spurious emissions can be suppressed, the in-band emissions may pass. They have to be reduced somewhere earlier within the transmitter chain, otherwise they may require a reduction in the RF output power, dropping the spectral efficiency of the system. In addition, high-order band pass filters, which might be required to suppress excessive out-of-band emissions, usually come along with larger in-band insertion losses. This could also decrease the overall power efficiency of the transmitter. Therefore, one should not put all the responsibility for the spectral purity of the emitted transmitter signal into the power amplifier and its matching network.

5.2.3   Up-conversion approach

The analysis of recent all-digital approaches reveals quite tough challenges related to the required clock in the GHz range. Therefore the (so-called) carrier-based approach is equally well established, avoiding really high frequency digital clocks. Instead, a local oscillator is used, typically embedded in a frequency synthesizer. The clear advantages are the better spectral purity of the local oscillator signal, easier and wider frequency tuning, and less stringent demands on the digital logic circuitry mainly determined by the minimum feature size of the available Si technology. Clear disadvantages are the necessity of RF components like coils and varactors, and usually larger silicon area occupation. Another potential drawback often discussed is the higher power consumption compared to more novel all-digital solutions. On the other hand, the analog design methodology is long established, guaranteeing the best performance and, furthermore, overall power dissipation is not necessarily so critical. Therefore, this carrier-based approach is attractive as well, and will be further investigated here. Figure 5.3 depicts its general principle.

Again, the main idea of this approach is the up-conversion of signals with well-defined pulse envelopes into the RF domain. In the literature, different methods of generating the pulses can be found (examples in [29, 73, 99]). Since the shape of the envelope substantially affects the final spectrum of the emitted pulses, special attention has to be paid to it. On the one hand, the regulatory limits given by the spectral mask must not be violated. On the other, as large an area as possible underneath the mask should be covered to achieve maximum output power and thus maximum transmission range. Here, many aspects may influence the decision for certain circuit realizations. For instance, employing an easy to implement ring oscillator for generating the carrier frequency may lead to a large variation of the center frequency of the pulses due to PVT variations.

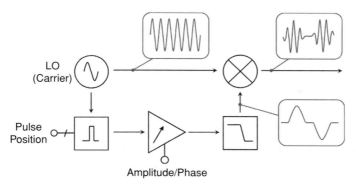

Amplitude/Phase

Figure 5.3 General principle of carrier-based pulse generation: up-conversion of the pulse envelope into the RF domain.

In order to prevent the violation of the mask, the pulse shaping has to ensure a narrow spectrum, always keeping the pulse emissions inside the mask. In case of further partitions of the allowed spectral frequency range into channels (as in IEEE 802.15.4a), a substantial variation of the pulse center frequency cannot be permitted. Instead, a more complex LO generation using an LC oscillator embedded in a phase-locked loop will allow a more precise filling of the mask even under typical PVT variations, albeit with a potentially higher power consumption. Here, only one of the possible implementations will be described in more detail. Nevertheless, the general considerations are also valid for other implementations and some simplifications might be possible, depending on the intended application. We will now return to Fig. 5.3, which represents a more complete implementation. In the upper branch, the LO signal (as previously mentioned, determining the center frequency of the emitted pulses) is multiplied or mixed with the envelope shape of the final pulses. In the case of an ideal multiplier, no unwanted harmonics appear within the output spectrum, whereas in the case of a typical up-conversion mixer, such mixing products appear and have to be dealt with. In literature, one can also find approaches with directly modulated oscillators, where the oscillation is started and stopped by the envelope signal [133]. The main problem here is that the time of oscillation is too short to control and to lock the frequency (for instance by a PLL). This can be overcome by either well-selected resonant components or by some kind of pre-calibration prior to pulse generation. The advantage of such an approach is the very limited power consumption which only takes place during the time it oscillates.

The lower branch in Fig. 5.3 belongs to the pulse envelope creation. Here, the LO also serves as timing for the pulse generation. This is very useful for any kind of phase modulation where a tight synchronization between LO signal and pulse envelope is required. Note that (quasi-random) phase modulation is reasonable, even in the case of non-coherent reception. By phase modulation of the pulses, their output spectrum becomes smooth instead of a comb-shaped structure. This allows for a better filling of the mask and thereby better spectral efficiency, which might be also the reason why phase modulation of the transmit signal is required in the standard IEEE 802.15.4a, even

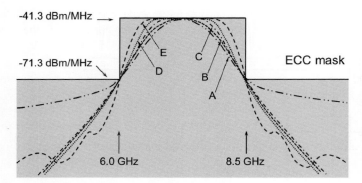

Figure 5.4 Resulting frequency spectrums of pulses (fitted into the ECC mask), generated by different pulse shaping filters: A – 5th order Bessel, 400 MHz; B – 5th order linear phase, 440 MHz; C – 5th order Butterworth, 640 MHz; D – 5th order Chebyshev, 780 MHz; E – 5th order elliptic, 950 MHz.

if simple energy detectors are used for reception. When using an integer-N phase locked loop for LO generation, one output of the divider chain can serve as the timing base for pulse generation.

By applying proper logic, a digital (rectangular) pulse can be generated representing the base of the later pulse envelope. It defines the time-of-transmission of the final pulse. As seen from earlier considerations of localization accuracy, the precision of this timing may need special attention. The rate and accuracy of the system clock determine the timing uncertainty of the transmitter. Therefore, depending on expected localization accuracy, the system clock has to be as high as necessary but will be at the cost of higher power consumption. Then, in the digital domain, any kind of pulse sequences can be generated and any kind of pulse position modulation can be applied based on that system clock.

The generated digital pulse has probably not got the proper shape to act as a final pulse envelope. Therefore, a pulse-shaping filter is inserted which combines the digital pulse with the transfer function of the filter. Since the duration of the digital pulse cannot be zero, its prior shape will also influence the emerging pulse envelope and finally the output spectrum of the pulses. Figure 5.4 shows a few examples where different pulse-shaping filters are stimulated with the same digital pulse and the emerged envelope is then up-converted into the RF domain using the same carrier signal of 7.25 GHz. All the filters were calculated in such a way that the resulting pulse spectrum fitted exactly into the ECC mask (−30 dB corners at 6.0 and 8.5 GHz). For filters A–E, we can observe an increasing coverage of the area underneath the spectral mask suggesting larger RF power. Unfortunately, this is accompanied by more ringing within the pulse shape (visible in the pulse response for the filters). Since the up-converted RF signal also includes such ringing, this could be detrimental in localization applications or in pulse-burst transmission schemes, where the ringing of previously transmitted pulses would be overlaid by following pulses.

As a result of the pulse-shaping filter, an analog signal has been formed which needs to be handled as such. One useful way to benefit from the analog nature of the signal at this point is to introduce possible amplitude and/or phase modulation here. While amplitude modulation is rarely used, the (binary) phase modulation is typically mandatory as described earlier. A variable gain amplifier (based on a Gilbert cell) can be used to vary the amplitude as well as the sign of the pulse envelope signal. Actually, one may also consider employing amplitude modulation to control the transmitter output power. In order to reduce the overall power consumption, this would be better applied directly to the power amplifier since this is where the largest amount of DC power could be saved.

5.3   Pulse detection

5.3.1   Bandwidth-related design trade-offs

The advantages of the very large signal bandwidth of UWB transmissions have already been discussed extensively. For instance, the resolvability of multipath signals significantly reduces the required fading margin, which is needed in typical narrowband systems. In order to benefit from that, the receiver needs to cope with the large signal bandwidth and maintain the included information within the received signal. Obviously, a fundamental trade-off between manageable signal bandwidth within the receiver and resulting power consumption arises, influencing the design decisions concerning the architecture. In addition, resolving and evaluating multipath signals takes a lot of computational power, which might be limited in mobile and/or battery-powered devices. Thus, the decisions regarding sampling rate and resolution will have an important impact on the achievable performance.

A second important aspect with regard to the large signal bandwidth is the achievable receiver sensitivity. Metaphorically speaking, a wide open receiver (in terms of bandwidth) opens the door not only for intended signals, but also for a wide variety of disturbing signals such as noise. It is known from theory that the spectral power density of thermal noise at room temperature is −174 dBm/Hz. As an example, defining the receiver bandwidth to be 1 GHz causes a noise floor of −84 dBm, which fundamentally limits the receiver sensitivity. Since the permitted transmit power of UWB signals is very limited, the receiver sensitivity becomes even more important in terms of the achievable radio link distance. Additional implementation losses like noise contributions of active devices (summarized by the noise figure) or insertion losses of matching networks, etc., will further lower the receiver sensitivity translating into a reduced communication range. Therefore the trade-off between receiver bandwidth and receiver sensitivity is another important design criteria in defining the UWB RF subsystem parameters, since it directly affects the localization accuracy, the possible raw data rate, the total power consumption, the communication range, and ultimately the cost of the system. In the following, simpler receiver architectures (non-coherent pulse detection) as well as more complex architectures (coherent pulse detection) are discussed; their eligibility depends on the application.

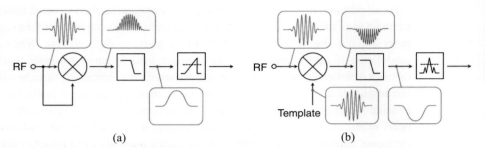

Figure 5.5 Pulse detection principles: (a) non-coherent energy detection – signal waveforms show all-positive signals after squaring; (b) coherent cross-correlation – signal waveforms suggest a perfect correlation between the received pulse and a template with opposite phase, showing conservation of phase information.

5.3.2   Detection concepts

How can one detect the short pulses with a duration of about 1 ns? Firstly a traditional narrowband receiver would only receive a small portion of the signal's energy which might result in too low an SNR. Secondly such a receiver would broaden the pulse in the time domain which reduces the accuracy of the time measurement. Therefore, a wideband receiver is needed which captures most of the transmitted energy of the UWB signal. The best way to transmit UWB signals are these very short pulses (in the time domain), which in fact provoke an equally short event at the receiver to be recognized and eventually memorized for further processing. Nothing is transmitted between two consecutive pulses, which could help the receiver to stay tuned and synchronized. But still, due to the wide receiver bandwidth, quite a lot of unwanted RF energy will be captured during the gaps which may be interpreted as pulse reception. Therefore, two main procedures have been established for short pulse detection: (i) energy detection within a given time interval (non-coherent detection), and (ii) the cross-correlation of the received signal with the expected pulse template (coherent detection). Both methods are illustrated in Fig. 5.5. The energy detection principle shown in (a) is typically based on squaring the received signal, integration of the captured energy within a short time interval (corresponding to the pulse duration), and evaluation of the integration result. The cross correlation principle shown in (b) is typically based on the continuous or periodical multiplication of the received signal with a template signal, summation over a correlation time interval, and evaluation of the result. For a more general description of modulation and coding as well as detection concepts, the reader is referred to Sections 2.6–2.8.

5.3.3   Energy detection principle

At first glance, the non-coherent energy detection looks easier to implement since there is no need to generate a template. Instead, the received RF signal will be squared in a multiplier. For an ideal multiplier with a gain of 1, the dynamic range of the multiplication product is doubled on a logarithmic scale referring to the applied RF signal. For instance, a typical dynamic range of 60 dB for the RF input signal would lead to a 120 dB dynamic range for the multiplication product. Such a dynamic range would be very challenging for the following stages. In addition, a real multiplier will always introduce some additional noise (denoted by its noise figure), resulting in a lower operational limit for the input signal. These two aspects lead to the conclusion that the received RF signal needs to be amplified (preferably by an LNA) and the gain should be variable to reduce the dynamic range of the intermediate signal adaptively. The more variable gain that is applied in the RF domain, the less are the requirements to the multiplier stage in terms of linearity and noise contribution. The most convenient situation for the multiplier would be to entirely absorb the dynamic range already within the RF domain by introducing proper RF variable gain amplifier (VGA) stages. Unfortunately, this causes quite high power consumption in the RF domain and, if the total gain is large (more than 50 dB), severe stability issues may arise in the RF amplification chain. On the other hand, leaving some variability in the RF signal prior to the multiplier will lead to a variable conversion gain of the multiplier depending on the strength of the input signal, which is actually impossible to predict. The automatic gain control of the receiver gets one more unknown parameter, making the algorithm more complicated. In [78] one can find an attempt to cope with this situation by splitting the total variable gain of the receiver amplification chain into two parts, one applied in the RF domain, the other in the baseband signal domain.

Assuming appropriate amplification and squaring of the RF signal, a low-pass filter has to be applied to remove unwanted multiplication products (such as twice the center frequency of the received pulses) and as much high-frequency noise as possible. Nevertheless, the in-band part of the noise as well as in-band interferers will remain within the signal. Since the signal squaring leaves only positive signal components, the uncorrelated noise no longer cancels out. This leads to a noise floor with a positive average value (DC component) which is typically strong, and is dependent on the gain setting of the amplification chain (due to its own noise contribution). Another drawback of the energy detection is the loss of phase or frequency information about the received signal. In [83] and [36], implementations of non-coherent receivers can be found.

Finally, in order to successfully detect an incoming pulse, it has to have a sufficiently larger amplitude than the noise plus interferer level at the decision stage (e.g. comparator). Here, a trade-off arises between the sensitivity of the detection versus the duration of the integration and sampling interval, which affects the receiver sampling rate and thus the power consumption of succeeding stages. To explain this, imagine a 1-bit decision circuit which requires a pulse amplitude twice as large as the RMS value of the incorporated noise for secure decisions. Then, the duration of the integration interval in relation to the duration of the single pulse roughly determines the required SNR. As an example, if the integration window is 10 times larger than the pulse duration, the energy of the single pulse is spread by a factor of 10 – and therefore requires an amplitude 10 times higher prior to the integrator. Using more advanced decision circuits with the help of analog–digital converters (ADCs) would, of course, relax the necessary SNR but increase complexity and power consumption. Nevertheless, a fundamental trade-off between the sampling rate and resolution, and the necessary SNR affecting the receiver sensitivity, remains. In general, a meaningful selection of the sampling rate could be somewhere between the chip rate (inverse of the duration of one pulse) down to the pulse repetition rate (inverse of the time distance between two pulses). Low data rate systems can shift the trade-off induced limits by using more than one transmitted pulse to contribute to the decision about one data bit. For instance, applying 10 pulses per bit would provide 10 dB of processing gain, which would compensate for a less complex hardware architecture.

5.3.4   Cross-correlation principle

In contrast to the energy detection case, the first stage of the cross-correlation detector (Fig. 5.5(b)) multiplies the received signal with a template of the expected pulse. In the ideal case, the template should perfectly match the expected waveform of the received pulse, including all implementation impairments and channel effects. Then, unwanted and non-matching signal components like noise and interferers would make much less of a contribution to the correlation result and the following bit decision. Therefore the receiver sensitivity can be significantly better compared to that of the energy detection principle – allowing a substantial extension of the communication range. Unfortunately, the template generation and the timing of its provision is rather complex and power consuming. The template generator could be similar or even identical to the pulse generation of the transmitter, and might be implemented in the transceiver architecture anyway.

The timing of the template generation in relation to the received signal is actually the main problem in this concept. In [174] this topic has been appropriately discussed. It is clear that an evaluable multiplication product appears only if the received pulse and the template fall together in time. Furthermore, if the phase information is to be derived, the necessary timing accuracy is only a fraction of the cycle duration of pulse center frequency. For instance, assuming BPSK to be recognized and assuming a center frequency of 8 GHz, the minimum required timing accuracy would be 31.25 ps (corresponding to a 90° phase shift). Unfortunately, the exact time of reception of a pulse cannot be known. Neither the time of emission at the transmitting node nor the time invariance of the channel can be predicted at the required accuracy. Therefore, the uncertainty of the time of reception is orders of magnitude larger than the required timing within the cross-correlation. In order to cope with this situation one can parallelize the correlation with a certain time delay between the parallel correlation branches. For a given time step size and a number of parallel correlation branches, a larger time window can be observed seamlessly. The longer the observation window, the shorter the synchronization procedure can be. For a perfect continuous cross-correlation in a circular fashion one would need as many parallel branches as the duration of the template divided by the time step size. The circuit complexity will depend heavily on the number of parallel branches, as illustrated in Fig. 5.6. One option to reduce it is a two-step correlation, where the first correlation recognizes the presence of a pulse using a larger step size, and the second evaluates the phase employing only a small time window using minimum step size.

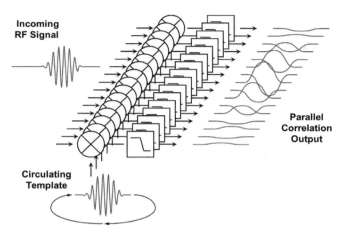

Incoming
RF Signal

Parallel
Correlation
Output

Circulating
Template

Figure 5.6 Illustration of the parallel cross-correlation with step-wise delayed templates.

A sliding correlation can be implemented either by stepwise delaying the received signal or by providing stepwise delayed templates. The first option is harder to implement, because the (still analog) received signal would become corrupted by feeding it through a delay element. Added noise as well as variation of the signal amplitude would increase through the chain, delay element by delay element, until the end, and the last correlation branch would produce a much worse result. In addition, the exact delay time of an analog delay element is affected by PVT variations making it hard to control and predict. The uncertainty from these variations becomes significant in longer chains of delay elements, making a continuous cross-correlation with a large number of parallel branches difficult to achieve. Therefore the opposite approach of generating stepwise delayed templates seems to be more feasible. For instance, taking a quantized and sampled representation of the template and letting it circulate in a shift register could provide the base of such a solution. Here the time step size and its accuracy are determined by the clock. In addition, all time delayed templates are identical in their shape, preventing any unwanted variations due to the number or position of the parallel correlation branches.

We will now continue with the more promising approach of providing delayed templates. As already described above, even the implementation of this concept is rather complex and requires exact timing between the parallel correlation branches. So how can we simplify the detection scheme without losing too much performance? There are three possible options.

1. The number of parallel correlation branches can be lowered, reducing the circuit complexity by the same factor. This comes at the expense of narrowing the observation time window, which finally affects the synchronization speed of the system. For instance, reducing the observation window by half would double the average synchronization time of the system. In addition, the danger of losing the synchronization increases. In cases of changing channel conditions (due to moving mobile nodes) the expected received signal may fall out of the observation window. Consequently, the

searching algorithm needs to be started from the beginning, which usually consumes quite a lot of time.

2. The complexity may be reduced by relaxing the stringent timing requirements between the correlation branches. Here, the phase recognition of the received signal is determined. If the phase of the pulse does not contain data to be demodulated, then the time step between the branches could be enlarged and its accuracy is weaker. Note that this would also reduce the ranging or localization accuracy in time-of-flight measurement-based applications.

3. The complexity of a cross-correlation receiver may be reduced by simplification of the template generation. Ideally, the best correlation result will be achieved if the template perfectly matches the received pulses. In the case of employing a DAC for template generation, one already faces imperfections by quantization errors in accordance with the implemented resolution. Nevertheless, even for a 1-bit resolution (rectangular template shape) the cross-correlation result is significant. If the performance loss due to the shape mismatch between template and received pulse can be tolerated, then a substantial simplification of the template generation is possible; it would be reduced to a gated oscillator (clock generator). Furthermore, trying to implement desirable continuous correlation would make even the gate obsolete. The entire receiver structure degenerates towards a traditional direct down-conversion architecture, where the correlation result behaves like the baseband signal with a comparatively large signal bandwidth. By applying a low-pass filter the pulse envelope can be extracted and further processed. Corresponding implementation examples can be found in [171, 189, 192]. The phase information of the received pulses in relation to the template (local clock) is still preserved and the usual quadrature direct down-conversion architecture could be employed to recover it. In the light of the discussion above this can be seen as two continuous cross-correlations with a time shift corresponding to a 90° phase shift (a quarter of the cycle time of the center frequency).

5.3.5   Transmitted reference scheme

For the sake of completeness, receivers for transmitted reference schemes should be mentioned. These receivers can be seen as a mixture of the two concepts above. The idea is to transmit two pulses instead of one, where the first acts as template (or reference) for the second. The two pulses are transmitted with a well-defined delay. On the receiver side the signal is fed into two paths, where one is delayed by the same amount of time as the duration between the reference and data pulses. Because of the delay in one signal path, the two pulses appear at the inputs of a multiplier producing an output signal at the same time. The detection fails if the time delay of the receiver does not match the duration of the two transmitted pulses. The main advantage of this concept is the elimination of the template generation. The received template matches very well to the data pulse since they face the same channel conditions. On the other hand, additions of noise and other disturbances to the template pulse degrade the correlation result. The main implementation issue is the delay element within one of the split paths of the

Table 5.1 General comparison of non-coherent (energy detection) and coherent (correlation) receivers.

	Non-coherent (energy detection)	Coherent (cross-correlation)
Receiver sensitivity	lower	better
Pulse detection probability	lower	better
Robustness against interferer	lower	better
Pulse phase recognition	impossible	possible
Bandwidth efficiency	lower	higher
Clock accuracy requirement	low	high
Synchronization speed	slow	fast
PLL requirement	low	high
Localization accuracy	lower	better
ADC requirement	low	high
Average power consumption	lower	higher
Circuit complexity	less	high

receiver. As already discussed above, the required accuracy (delay time, signal shape distortion) is hard to achieve in monolithic integrated circuits. These delay elements would need to be tuned in accordance with the LO drift of the transmitter and any kind of PVT variations have to be canceled out. In fact, the receiver complexity increases due to these implementation requirements while the receiver sensitivity improvements might only be moderate.

5.3.6 Comparison

The decision about the receiver concept at the beginning of a design project will have significant implications during the implementation, and the important advantages and disadvantages are summarized in Table 5.1. Not included in the comparison are receivers for transmitted reference transmission schemes, since their performance and complexity lies somewhere between non-coherent (energy detection) receivers and coherent (correlation) receivers.

5.4 RF frontend components

Many of the components in an IR-UWB transceiver are identical or at least similar to those in traditional narrowband radio systems, except the fact that the signal bandwidth is much wider. Descriptions and design strategies can be found in many good textbooks. However, in particular, the RF front-end consisting of the low noise amplifier on the receiver side and the power amplifier on the transmitter side have to be different due to the lack of successful narrowband impedance matching techniques. Here, a compromise between broadband power matching, noise optimum matching and power dissipation is needed. In addition, stability issues and phase linearity issues arise as a consequence of

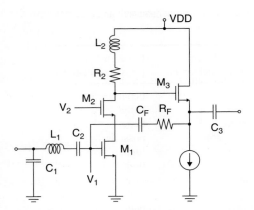

Figure 5.7 Simplified schematic of a two-stage LNA with series shunt feedback, first stage with cascode topology and buffer output stage.

the broadband behavior of the active components. Therefore these two components will be discussed in more detail.

5.4.1   Low-noise amplifier

We will start with the low-noise amplifier. Its task is to provide sufficient gain to overcome the noise contributions of the following stages without adding too much of its own noise to the received signal. In addition, the input should be power matched as perfectly as possible to the source to ensure maximum RF signal transfer from the antenna to the amplifier. Note that optional antenna switches and band-pass filters may also need to be involved in this consideration. Furthermore, a high and flat broadband gain is required while ensuring sufficient linearity and low power consumption. Unfortunately, there is no perfect all-in-one solution, leading to many different approaches tackling different difficulties such as differing broadband power and noise optimum matching. Nowadays, one approach can be found quite often: a 2-stage amplifier with series shunt feedback, where the first stage is built as a cascode or folded-cascode stage. One example, including a design description, can be found in [37]. A simplified schematic is shown in Fig. 5.7.

The feedback is used as a good compromise between input impedance matching and overall noise figure. In contrast to this, the most appropriate narrowband technique using inductive (source/emitter) degeneration does not allow satisfying broadband operation (flat gain and noise performance). Also, the often-used common-gate topology as an input stage has some drawbacks with respect to the noise performance, since the required biasing is usually too far away from the noise optimum. Nevertheless, this topology has a natural advantage concerning the electrostatic discharge (ESD) protection of the input.

Using a resistive feedback obviously lowers the overall gain, but due to the better wideband input impedance matching the noise figure is also reduced. The cascode topology of the first stage helps to enlarge the gain and thereby widens the frequency range of operation. If the frequency band of operation is not too large, one can consider inductive peaking at the load. Depending on the quality factor of the coil, the gain can be enhanced while pinching the bandwidth. This means the maximum achievable gain and the bandwidth have to be traded against each other, depending on the intended application. For instance, a channel-selective inductive peaking seems to be the optimum approach regarding the requirements of relatively small channels defined by the standard IEEE 802.15.4a. In addition, using inductors as load elements could also be considered to improve the gain flatness in the case of negative gradient gain over frequency. Here, it is important that the self-resonance frequency of the coil is above the operational frequency – to obtain gain improvement over frequency. Achieving large inductor values and high quality factors at the same time could be challenging in today's silicon technologies.

The two-stage approach for the LNA permits a more independent optimization of the input and output stages. This allows a better optimization of the cascode stage in terms of noise figure. The second stage can be more easily adapted to the succeeding stage. In addition, the reverse isolation is further improved. Obviously, the improvements by the use of two stages come at the expense of potentially larger power consumption. While the feedback capacitance is simply used to separate the biasing at the input and output, the feedback resistor determines the gain reduction and the input impedance matching improvement. On the other hand, it does not significantly affect the noise figure of the LNA. Since the resistor cannot be too large (to avoid stability problems), one could also insert an inductor into the feedback. This would (of course) introduce a frequency dependency leading to increasing gain at higher frequencies. One could benefit from this as well, as compensation for the usual gain deficits over frequency.

5.4.2   Power amplifier

The other important front-end component is the power amplifier of the transmitter. The power consumption of the PA can easily dominate the whole transmitter. It has to deliver a certain peak power to the antenna usually representing $50\,\Omega$ impedance. For instance, expecting a peak-to-peak voltage of 1 Vpp at the antenna would require a temporary RF current of 20 mA, which is already quite a large value in integrated RF circuits. On the other hand, this RF power capability is demanded only during real pulse transmission, whereas during the gaps between two consecutive pulses no RF power needs to be delivered. Since typical duty factors might be 1% or even below, one can imagine that a traditional class-A power amplifier would become really power inefficient. Ideally, the PA should consume DC power only at the time of a momentary pulse transmission. Unfortunately, the deployed power devices have to withstand the quite large peak power even if the duty cycle is very low. This results in large device dimensions bringing along large parasitic capacitances. In fact, they slow down the essential switching speed needed to switch the PA circuit on and off in power-saving modes. Subsequent necessary preparation time intervals of the PA before the actual pulse transmission can be much longer than the duration of a single pulse itself. In addition, fast power devices in modern Si technologies usually suffer from limited breakdown voltages requiring careful considerations of the intended pulse peak power. Therefore, a clever co-design of the transmission scheme (determining PRF, pulse peak power, symbol and frame rates, etc.) together with the RF front-end circuit design (influenced

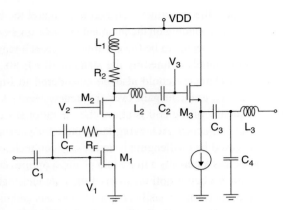

Figure 5.8 Simplified schematic of a two-stage PA with series shunt feedback, first stage with cascode topology and buffer output stage.

by device cut-off frequencies, breakdown voltages, current driving capabilities, etc.) is mandatory to minimize the overall average power consumption of the system.

More digital-like approaches to implement a PA have already been mentioned above. They are very promising in terms of power efficiency if difficulties concerning the spectral efficiency (coverage of the available spectral mask without violating the mask) of the generated RF signal could be overcome. A number of recent publications concerning this topic can be found in the literature. Another quite obvious approach would be to take a class-A PA circuit topology, design it for broadband operation, and add fast switch-on and switch-off capabilities to improve the average power efficiency. Such a design is described in the following.

Similar to the depicted LNA design, the PA could also be composed of two stages, allowing more independent and flexible optimization of the input and output stages. Again, the first stage is a cascode stage providing large gain, whereas the input matching is constructed in conjunction with the previous stage. The second stage is basically just a driver (or buffer) to provide the intended RF power to be delivered to the output impedance (usually 50 Ω). The employed techniques to obtain broadband impedance matching and flat gain characteristic over frequency are similar to the LNA design. A resistive shunt feedback is used to enlarge the bandwidth and flatten the gain. Inductors as load elements are employed to increase the gain on higher frequencies and they might also be used to enlarge the gain at the expense of narrowing the bandwidth. One example, including a design description, can be found in [112]. A simplified schematic is shown in Fig. 5.8.

So far, this topology is not very power efficient. This disadvantage could be overcome by introducing a fast switchable biasing circuitry which reduces the DC power consumption to almost zero if no pulse has to be transmitted. Fortunately, the transmit scheme is well defined and can easily be applied to the PA on-chip. Note that switching the PA supply on and off would be far too slow (of the order of microseconds to milliseconds) compared to the duration of a pulse (about 1 ns). The reason for this are the large blocking capacitances (plus large parasitic capacitances), which are used to steady the on-chip supply voltages. After the recharge, currents will be limited to a certain level – hence the recharge duration grows. Ideally, the switching delay should be of the same order of magnitude as the pulses themselves to minimize the wasted DC power during the gaps between two consecutive pulses. Probably the best way to do this is to switch the biasing of the main DC power consumers. In most cases this will be the final stage of the PA. Depending on the topology of this stage, one could switch the tail current in differential stages, or the base current of bipolar junction transistors (BJTs), or the gate voltage of metal oxide semiconductor field-effect transistors (MOSFETs). The design of such a bias switching option has to be done very carefully since biasing nodes are naturally very sensitive to any disturbances like noise. Therefore, large blocking capacitances are usually employed to provide a good AC ground. Unfortunately, they would contradict the desire to introduce fast bias switching. Another option is the usage of inductors within the biasing network, but they require a large chip area. Thus, like in most cases, the design preferences in terms of circuit performance, chip cost (silicon area), overall power consumption, and demanded level of monolithic integration, will influence the design decisions.

5.5   Monolithic integration

In general, the monolithic integration of the RF subsystem of any wireless communication system is very desirable in terms of form factor, power consumption, and final system cost. Compared to a discrete solution with off-the-shelf components, for large volumes the development costs might be justified by further reduced production costs. It is clear that the huge signal bandwidth of UWB systems introduces new circuit design challenges. Traditional RF design techniques for narrowband systems are no longer applicable in the usual way. Nevertheless, it becomes even more important to match the entire RF design to the expected frequency range of operation. This improves the DC power efficiency of the system and reduces the susceptibility to out-of-band interferers and noise. On the other hand, maintaining the wide signal bandwidth at interstage interface connections seems to be much easier on-chip in a monolithic IC than on a PCB with discrete components. Therefore IR-UWB radio circuits are grateful candidates for a high level of integration.

As for any new radio design, the question arises whether to target a single-ended or a differential design. The well-known advantages of differential designs are lower common-mode noise or interferer generation and higher tolerance to ground inductance. In particular, ground bond wire inductances do substantially degenerate the gain and degrade the input impedance matching in many cases. Therefore differential design improves the immunity against supply and substrate noise. On the other hand, the power dissipation is usually doubled for a given cut-off frequency of the circuit. Since UWB circuits are quite sensitive to any kind of disturbances due to their wide bandwidth, a fully differential design is always preferable even at the expense of higher power consumption and a larger silicon area. The only obstacles are usually the external RF components such as antennas, switches and band-pass filters, since many of these devices are designed for single-ended operation. Furthermore, the broadband conversion between single-ended

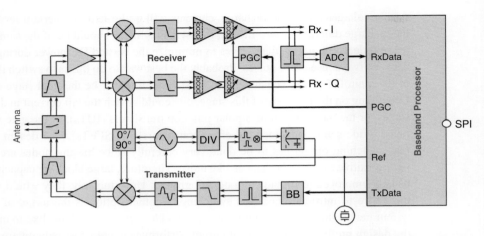

Figure 5.9 Block diagram of an IR-UWB transceiver subsystem, consisting of the entire RF and analog front-end, ADC and baseband processor.

Figure 5.10 Die photograph of a fully monolithic integrated IR-UWB transceiver, intended for compliant operation with the standard IEEE 802.15.4a, fabricated in a 0.25 μm SiGe-BiCMOS technology of IHP (Germany).

and differential signals (and vice versa) is quite difficult, since in general the time domain pulse shape will not be distorted. Nevertheless, there is a clear tendency towards fully differential designs. Figure 5.9 shows the block diagram of a fully integrated and fully differential IR-UWB transceiver to illustrate architectural considerations. Note that many other transceiver architectures are also possible.

Table 5.2 Key performance parameters of the IEEE 802.15.4a compliant single chip transceiver.

Parameters	Performance
Si technology	IHP SGB25V
Chip size	3.25 μm × 3.25 μm
Operation channel	7.9872 GHz
Data rate	0.85 Mbps
Signal bandwidth	500 MHz
Reference clock	31.2 MHz
PLL settling time	<10 μs
Baseband processor interface	SPI
Baseband processor clock	31.2 MHz
Transmitter RF output amplitude	250 mVpp
Transmitter modulation schemes	BPSK, OOK
Receiver sensitivity	−65 dBm
Receiver maximum voltage gain	75 dB
Programmable gain setting	4 bit
ADC resolution	6 bit
Supply voltage	2.85–3.3 V
Power consumption – BB processor	0–44.0 mW
Power consumption – frequency synthesizer	1–68.5 mW
Power consumption – transmitter	1–65.5 mW
Power consumption – receiver	1–65.0 mW
Power consumption – ADC	1–70 mW
Sleep modes switching delay	<2.5 μs

Not shown in the block diagram are auxiliary units for supplying the main RF parts. Nevertheless, they are quite important because they provide some isolation concerning supply noise and, in addition, introduce the opportunity of sleep modes. Depending on the transmission and reception situation, the baseband processor can decide to switch on and off entire RF parts (e.g. receiver, transmitter, frequency synthesizer). This feature is largely used by duty-cycling schemes, especially in wireless sensor networks where the average power consumption becomes really important. Here, the lower limit is determined by the leakage power consumed by the sleeping components and the required guard times for proper operation. As mentioned in the scope of a PA design, reducing the margin for these guard times is always a design goal, thus allowing the extension of the battery life of the mobile device. Therefore, consideration of how and when components of the RF subsystem can be switched should be done at the beginning of the design project.

One example implementation of an IR-UWB transceiver in accordance with the block diagram in Fig. 5.9 is shown in Fig. 5.10. It is intended for compliant operation with the standard IEEE 802.15.4a in the high band channel (7.9872 GHz) for worldwide operation. It has been fabricated in a 0.25 μm SiGe-BiCMOS technology of IHP GmbH [69]. The chip size is about 3.25 × 3.25 μm. Table 5.2 summarizes the key parameters of this circuit.

6   UWB applications

Xuyang Li, Jens Timmermann, Werner Wiesbeck and Łukasz Żwireło

In this chapter typical UWB applications are presented and discussed, particularly regarding their RF system design.

6.1   UWB communication

6.1.1   Modeling of the system components

The system model for impulse radio transmission consists of the transmitter, the UWB indoor channel and the receiver, see Fig. 6.1. At the transmitter side, the pulse generator creates pulses at moments defined by the pulse position modulation (PPM) scheme and the time-hopping (TH) code. Without modulation and coding, pulses appear at multiples of the pulse repetition time T. To modulate the signal by PPM, the binary values of the bit stream determine whether a constant delay T_{PPM} is introduced or not with respect to the pulse repetition time.

The TH code adds a further pseudo-random delay to smooth the spectrum. This code must also be known at the receiver to demodulate the signal. To ensure that the radiated signal does not violate the spectral mask, an analog transmit bandpass filter is placed before the transmit antenna. The radiated signal propagates via an indoor radio channel including AWGN towards the receive antenna. It is then amplified, filtered and demodulated; different demodulation schemes are available. The following subsections describe the modeling of the non-ideal system components.

Pulse shape generator

Classical considerations emphasize the idea that impulse radio is a low-cost technique because of its simplified hardware architecture. For example, there is no necessity to use up-converters, since signals are radiated in the baseband. To remain low cost, the associated pulse shapes result from low complexity devices. The problem is, however, that classical pulse shapes often violate the spectral mask and do not fully exploit it, which leads to a reduced signal-to-noise ratio (SNR) and to a reduced performance. In applications where the performance is relevant but not the cost, optimized pulse shapes help to improve the SNR. They can be obtained by pulse shaping networks [43]. The efficiency η of a pulse with respect to the FCC regulation can be calculated as shown

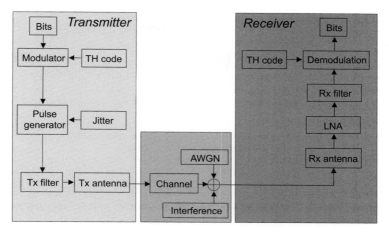

Figure 6.1 System model for non-ideal impulse radio transmission. ©2010 KIT Scientific Publishing; reprinted with permission from [168].

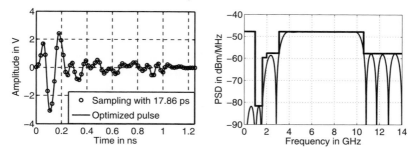

Figure 6.2 Optimized pulse shape, left: in the time domain; right: in the frequency domain. ©2010 KIT Scientific Publishing; reprinted with permission from [168].

in (2.32). The present contribution uses an optimized pulse shape with an efficiency of 90% (see Fig. 6.2).

In the system simulator, the pulse shape is modeled in the TD by samples. The sampling time step is $T_0 = 17.86$ ps, which ensures sufficient resolution to describe the pulse shape in the TD. Ideally, the pulse shape generator creates pulses at multiples of the pulse repetition time. Since the oscillator driving the pulse shape generator may be affected by jitter, pseudo-random positive and negative delays with respect to the ideal clock position are introduced to simulate the jitter. The probability density function of the jitter delay is modeled by a Gaussian behavior with a mean value of zero and a standard deviation σ_{jitter}.

6.1.2   Modulation and coding

The present contribution concentrates on PPM. This modulation scheme causes the pulse shape to be shifted in time by T_{PPM}, or it remains unshifted, depending on the

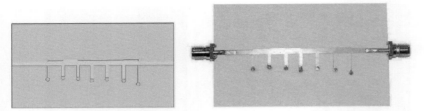

Figure 6.3 Analog filter for the FCC regulation: CAD and fabricated models. ©2010 KIT Scientific Publishing; reprinted with permission from [168].

actual bit value. A PPM modulated signal shows strong periodicities in the TD, which cause high discrete peaks in the FD. These peaks can violate the spectral mask and must somehow be attenuated. Coding is a suitable solution which may be used anyway both to separate users and to smooth the spectrum. The system model uses a two-stage TH code. First, the pulse repetition time is separated into $N_{TH,1}$ time slots. The length of a time slot is $T_{TH,1} = T/N_{TH,1}$ which should be a multiple of the time step T_0 because the system simulator performs a discrete simulation. The pseudo-random TH determines a slot number and shifts the pulse shape into that slot. Then, a second TH code determines an additional fine time shift as multiples n of T_0 that yields the condition $n \cdot T_0 < T_{TH,1} - T_{PPM} - T_p$. Here, T_p is the duration of a pulse. The condition ensures that there is no overlap of the pulse shape with the next time slot. For the simulation, the following values are taken: $N_{TH,1} = 4$ and $n = 31$. The fine TH code can be seen as a dither code [181], which is mainly responsible for smoothing the spectrum.

Filter

To ensure that the radiated signal always – and independently of the pulse shape – fulfills the FCC mask, an analog transmit filter must be included. The desired transmission behavior corresponds to the FCC regulation. One way to design an ultra-wideband filter of order N_F using the microstrip technique is to use N_F short-circuited stub lines of one-quarter wavelength that are connected by one-quarter wavelength transmission lines [136], whereas the midband wavelength is used in the design process. Increasing the filter order results in a steeper filter slope; this technique is used to design a filter for the FCC mask. To take into account the non-ideal filter slope in the filter design process, the chosen passband frequencies are not 3.1 and 10.6 GHz, but 3.5 and 10.2 GHz. The chosen filter order is $N_F = 7$, and Chebyshev filter coefficients with 0.1 dB ringing inside the passband are taken. By applying the filter design equations from [136], the widths and lengths of the microstrip lines can be determined. After optimizing the filter using CST Microwave Studio 2008 simulation software with respect to the FCC mask, the final filter is fabricated and shown in Fig. 6.3.

Figure 6.4 demonstrates its transmission and reflection behavior for both the simulation and the measurement. It can be seen that the simulation and measurement agree

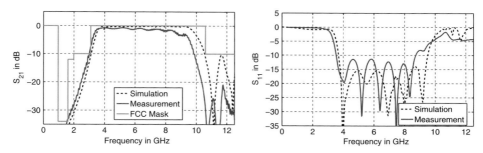

Figure 6.4 Transmission properties of an FCC filter. ©2010 KIT Scientific Publishing; reprinted with permission from [168].

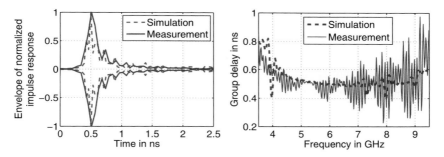

Figure 6.5 Envelope of normalized impulse response and group delay of an FCC filter. ©2010 KIT Scientific Publishing; reprinted with permission from [168].

well and that reflections are below −10 dB inside the relevant frequency range. The envelope of the impulse response and the group delay behavior are shown in Fig. 6.5 for both the simulation and the measurement. From the envelope of the impulse response a ringing of approximately 0.9 ns can be determined for $\alpha = 0.1$. The measured group delay variation is of the order of 0.8 ns. Group delay variations versus frequency can result in inter-symbol interference since the pulses are spread in the time domain. As a consequence, system design parameters that directly influence the temporal properties of the signal – such as the TH slot length, the PPM offset and the pulse repetition time – must always be adapted with respect to the group delay behavior of the system components in order to minimize the distortions. The measured S-parameters of the filter describe the filter model, which is part of the system simulator. Measurement data is available between 2 and 14 GHz at 801 equidistant frequencies. Since a system simulation with a time step of 17.86 ps needs data from 0 to 28 GHz, the remaining frequency data is described by the S-parameters of an ideal bandpass behavior, where the frequency step is not changed. The filter model is used both at the transmitter (Tx) side and at the receiver (Rx) side to describe the Tx and Rx analog filters.

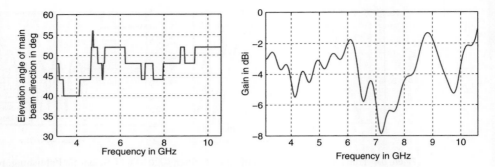

Figure 6.6 Left: frequency dependence of the main beam direction of the antenna; right: frequency dependence of the gain at an elevation angle of 90°. Left: ©2010 KIT Scientific Publishing; reprinted with permission from [168]; right: ©2009 De Gruyter; reprinted with permission from [169].

Antennas

Besides a good match and the desired radiation pattern in the whole frequency band of interest, UWB antennas should have a low signal distortion (see Section 2.3.4 on Ringing). A typical UWB antenna with a nearly omnidirectional azimuth pattern inside a large frequency range is the monocone antenna, as shown in Fig. 2.8. This figure also illustrates the elevation angle θ. At a given frequency, the three-dimensional (3D) radiation pattern of this antenna has been measured with an angular step of 4° in an anechoic chamber. Measurement data is available at equidistant frequencies between 2.5 and 12.5 GHz with a frequency step of 6.25 MHz. The measured antenna gain versus elevation angle is shown in Fig. 2.8 (right) for three different frequencies. It can be seen that the maximum gain is at an elevation angle of about 50°. The exact angle associated with the maximum gain depends on the frequency, as shown in Fig. 6.6. It has to be mentioned that the angular resolution in Fig. 6.6 (left) is limited to 4° as a result of the measurement setup. The maximum gain $G(f)$ also depends on the frequency. Fig. 2.8 shows values for $G(f)$ of the order of 6 dBi. Since scenarios are investigated where the Tx and the Rx antennas are placed at the same height, the direct line of sight (LOS) propagation path is at $\theta = 90°$. Consequently, the behavior of the antenna gain at this elevation angle is of special interest and is shown in Fig. 6.6 (right).

It can be seen that the gain varies between -8 and -1 dBi versus frequency, which leads to a pulse distortion in the LOS direction. In the system simulator, the antenna model includes data from 0 to 28 GHz, whereas measurement data is available from 2.5 to 12.5 GHz; zero-padded data is taken for the remaining frequency range. The measured data describes the complex elevation patterns of the monocone antenna at 801 equidistant frequencies between 2.5 and 12.5 GHz with an angular resolution of 4°. The associated azimuth patterns are described by an omnidirectional behavior.

Indoor channel

The ray-tracing based indoor channel is implemented in the system model, as described in Section 2.2.

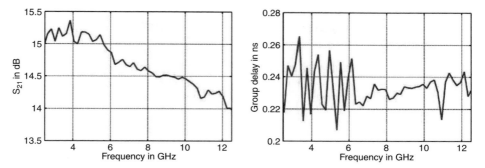

Figure 6.7 Left: transmission behavior of LNA; right: associated group delay. ©2010 KIT Scientific Publishing; reprinted with permission from [168].

Noise and interference

Noise and interference determine the SNR at the receiver side. For now additive white Gaussian noise (AWGN) is assumed. Since the antennas are located in an indoor environment, the received thermal noise can be modeled at room temperature. The associated noise power N is determined by

$$N = \sigma_\mathrm{n}^2 = kTB \tag{6.1}$$

where k is the Boltzmann constant, T the absolute temperature, and B the bandwidth. The expression kT represents the spectral noise density. For $T = 300\,\mathrm{K}$, $kT = -113.83\,\mathrm{dBm/MHz}$. A bandwidth of $(10.6 - 3.1)\,\mathrm{GHz} = 7.5\,\mathrm{GHz}$ delivers a total noise power of $-75.08\,\mathrm{dBm}$ in the UWB frequency band. A possible AWGN interference is modeled by an increased value of T.

Low noise amplifier

The LNA is the first element after the Rx antenna and filter at the receiver side. Its noise figure should be very small and its gain high since it dominates the noise figure of the whole receiver frontend. In the example presented here, the commercially available component HMC-C022 is used. Fig. 6.7 (left) demonstrates the gain behavior versus frequency S_{21}, which is rather flat inside 3.1 and 10.6 GHz with a variation below 1 dB. The associated group delay is visualized in Fig. 6.7 (right) and shows variations of the order of only 0.04 ns. To model the LNA, the full S-parameters of the data sheet [65] are taken into account as well as the 1 dB compression point and the 3rd order intercept point. The joint noise figure of the LNA and the following receiver frontend is modeled as 2 dB.

Coherent demodulation

The principle of coherent demodulation has already been described in Section 2.9.1. In the system simulator, coherent demodulation is achieved by an ideal multiplier and an ideal integrator that sum up the discretized values in the time domain. The template signal is based on the optimized pulse shape from Fig. 6.2. In the following, only PPM is considered.

6.1.3   Performance for different system settings – system analysis

The previous section showed in detail the modeling of non-ideal components for impulse radio transmission as well as the implemented coherent and incoherent demodulation methods. The interfaces can be realized in dedicated simulation programs such as the Ptolemy environment in Agilent's Advanced Design System (ADS) [15], so that flexible system considerations and simulations are possible. The following section visualizes the effects of non-ideal components and investigates the system performance in terms of bit error rates for various system configurations. The presented results are based on PPM and a two-stage TH code.

Visualization of the non-ideal effects

To study the effects of the single components, the following two subsections visualize the signal after each stage of the non-ideal system in both the FD and the TD. The investigated distance is $r_{TxRx} = 3.57$ m. Jitter effects are turned off, while all the other non-ideal components are taken into account.

The chosen system settings are:

$$T = 1600 \cdot T_0 = 28.57 \, \text{ns} \qquad \text{and} \qquad T_{PPM} = 200 \cdot T_0 = 3.57 \, \text{ns}.$$

The amplitude of the pulse from Fig. 6.2 (efficiency $= 0.9$) is set such that the pulse just meets the FCC regulation. Precisely expressed, the Tx antenna has a maximum gain of 6.4 dBi, which forces the pulse shape before the antenna to meet the limit of $-41.3 - 6.4 = -47.7$ dBm/MHz. The amplitude of this pulse also depends on the pulse repetition time T. It is tuned so that the PSD of the pulse at an observation period of T delivers a maximum value of -47.7 dBm/MHz. The power P_{dBm} in dBm of the signal between 3.1 and 10.6 GHz can be calculated as

$$P_{dBm} = 10 \cdot \log \left(\frac{\eta \cdot 7500 \, \text{MHz} \cdot 10^{-\frac{47.7}{10}} \frac{mW}{MHz}}{1 \, \text{mW}} \right) = -9.4. \qquad (6.2)$$

Non-ideal effects in the frequency domain

The following section visualizes the simulated PSD after each system component. Figure 6.8 visualizes the PSD of the PPM modulated signal. Integrating the PSD delivers a total power of -9.4 dBm. The spectrum is not flat and exceeds -41.3 dBm/MHz. However, the spectral characteristics change rapidly when TH coding is introduced. The coarse TH code suppresses many spectral lines while an additional fine TH code smooths the spectrum radically (see Fig. 6.9). The total power is unchanged at -9.4 dBm. A fine TH code can be seen as a classical dither code, which is used to smooth the spectrum of a signal. The maximum PSD value is now much closer to the desired -47.7 dBm/MHz. Further improvements of the smoothing can be achieved by an optimization of the maximum code length and a better TH code. Here, a primitive

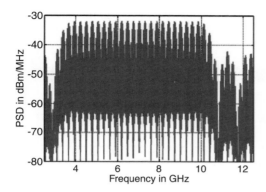

Figure 6.8 PSD of PPM modulated signal. ©2010 KIT Scientific Publishing; reprinted with permission from [168].

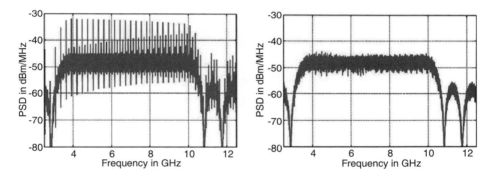

Figure 6.9 Left: PSD after coarse TH coding; right: PSD after additional fine TH coding. ©2010 KIT Scientific Publishing; reprinted with permission from [168].

polynomial is used to generate a TH code. Further possibilities are to use Gold code sequences, modulo operations, etc. [111].

Fig. 6.10 (left) shows the PSD after the Tx filter. It can be seen that the spectrum is still quite flat, but drops down slightly versus frequency due to the increasing insertion loss of the filter.

The power of the signal is $P = -11.3$ dBm which means that the filter introduces a mean attenuation of 1.9 dB. Due to the fact that the Tx antenna, the channel and the Rx antenna are modeled together by a single transfer function, the next signal that can be visualized is the signal after the Rx antenna. Its PSD is strongly distorted (see Fig. 6.10 (right)). The principal behavior is related to Fig. 2.10 (right) since the input spectrum at the Tx antenna, described by Fig. 6.10 (left), is quite flat. The power after the Rx antenna is $P = -72.14$ dBm which corresponds to an attenuation of 60.8 dB by the channel, including antennas. This value can be explained as follows. At a distance of $r_{TxRx} = 3.57$ m, Fig. 2.11 (right) presents about 55 dB attenuation for the ray tracing model (excluding antenna effects). The remaining attenuation of 6 dB results from the

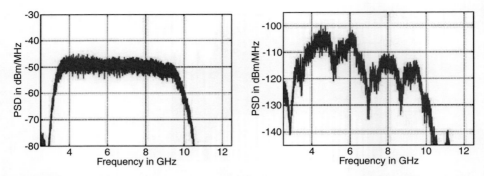

Figure 6.10 Left: PSD after Tx filter; right: PSD after the ensemble of Tx antenna, channel and Rx antenna. ©2010 KIT Scientific Publishing; reprinted with permission from [168].

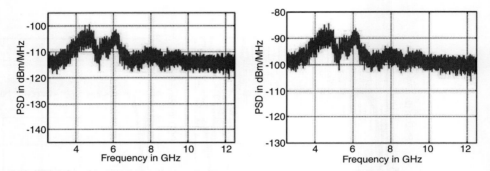

Figure 6.11 Left: PSD after thermal noise; right: PSD after LNA. ©2010 KIT Scientific Publishing; reprinted with permission from [168].

negative antenna gain which is about -3 dBi for both the Tx and Rx antennas in the LOS direction (see Fig. 6.6 (right)).

Figure 6.11 (left) shows the PSD after additive white gaussian noise, which delivers at least a level of -113.83 dBm/MHz. The PSD after the LNA is shown in Fig. 6.11 (right). Apart from an amplification of about 14 dB, the principal behavior is almost unchanged since the amplifier has more or less a constant gain versus frequency. The power after the LNA is -55.4 dBm. Finally, the Rx filter cuts out the relevant frequency range and delivers the PSD shown in Fig. 6.12. The power after the Rx filter is -56.8 dBm which means that the Rx filter introduces a mean attenuation of 1.4 dB. This attenuation differs slightly from the attenuation caused by the same physical Tx filter, which is 1.9 dB. This can be explained by the fact that the signal after the LNA has strong signal components only for frequencies lower than 6 GHz; below this frequency the insertion loss of the filter is still small (see Fig. 6.4 (left)).

Non-ideal effects in the time domain
The influence of the system components can also be shown in the TD. All hardware components add delays in the TD because of their physical length, and their non-ideal

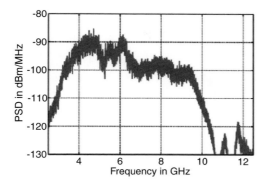

Figure 6.12 PSD after Rx filter (before demodulation). ©2010 KIT Scientific Publishing; reprinted with permission from [168].

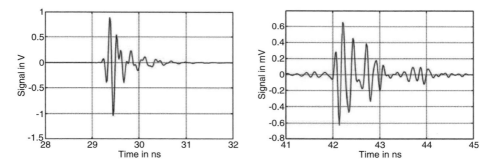

Figure 6.13 Left: pulse shape after FCC transmit filter; right: pulse shape after Rx antenna. ©2010 KIT Scientific Publishing; reprinted with permission from [168].

component behavior leads to distortions of the pulse shape. The following investigations visualize the distortion of the pulse shape of a PPM modulated signal using TH coding. The investigation aims at demonstrating the principal influence of the non-ideal system components.

The input pulse shape is visualized in Fig. 6.2 (left). Modulation and TH coding do not change the pulse shape, since only possible delays are introduced. The first component that distorts the pulse shape is the analog Tx filter. The distorted pulse shape is shown in Fig. 6.13 (left). The ringing of the pulse is increased due to the non-flat spectrum of the filter. A severe further pulse distortion is however caused by the channel, including antenna effects, which can be seen in Fig. 6.13 (right). The resulting pulse shape shows additional zero crossings and an increased duration. Fig. 6.14 visualizes the pulse shape after AWGN and after the LNA. The pulse shape after the final analog Rx filter is shown in Fig. 6.15 (left). The filtering process cuts out the relevant bandwidth. In the time domain, the resulting signal looks similar when compared to the synchronized template signal. Compared to the input pulse shape of Fig. 6.2 (left), the signal is strongly distorted and attenuated. Fig. 6.15 (right) shows the first part (positive pulse only) of

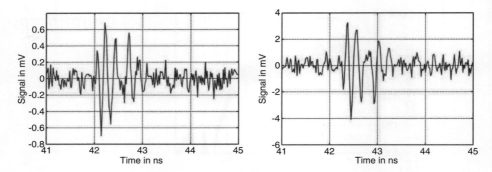

Figure 6.14 Left: pulse shape after AWGN; right: pulse shape after LNA. ©2010 KIT Scientific Publishing; reprinted with permission from [168].

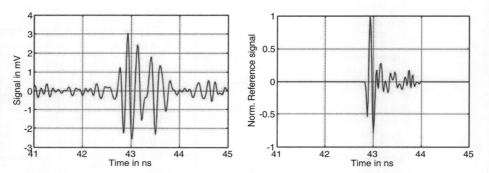

Figure 6.15 Left: pulse shape after Rx filter; right: synchronized reference signal (positive pulse only). ©2010 KIT Scientific Publishing; reprinted with permission from [168].

the synchronized template signal at the Rx side (recall that PPM demodulation makes use of a positive and a negative pulse). The template pulse is correlated (multiplication) with the received pulse shape from Fig. 6.15 (left). It can be seen that the main peak of Fig. 6.15 (left) and the main peak of Fig. 6.15 (right) are located at the same position (about 42.9 ns). The position of the second strongest peak also agrees and appears at 43 ns. Both signals are correlated sufficiently for a coherent demodulation with an acceptable performance. The performance in terms of bit error rates is analyzed in the next section. Comparing the original transmitted pulse shape with the one at the end of the receiver, it can clearly be seen that the overall pulse length (e.g. defined by the ringing) is drastically increased. This effect limits the usable pulse repetition frequency and data rate without intersymbol interference.

6.1.4   Performance for coherent demodulation

The system simulator allows for flexible system simulations based on different system configurations. The following sections present some chosen results on the achievable performance when an optimized pulse shape is used. The investigations concentrate on

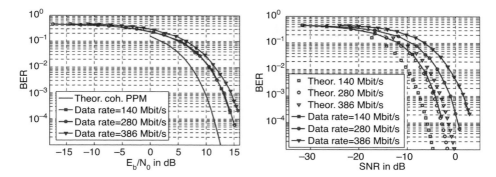

Figure 6.16 Left: BER vs E_b/N_0; right: BER vs SNR. ©2010 KIT Scientific Publishing; reprinted with permission from [168].

the influence of the data rate, a possible synchronization error, and system jitter for coherent demodulation. Finally, the performance is compared to the case of incoherent demodulation. In order to study only the effect of the non-ideal system components (and not the combined effect of hardware and coding), the double-stage TH code is switched off. The other system settings remain unchanged.

Influence of data rate

Figure 6.16 shows the bit error rate versus E_b/N_0 for different data rates R. The data rate $R = 280$ Mbit/s corresponds to the system settings presented in the previous section (e.g. pulse repetition time 28.57 ns). A higher (smaller) data rate is obtained by decreasing (increasing) the pulse repetition time. In the figure, solid curves result from simulations.

The graphs shown present degradations compared to the theory of AWGN channels [141], which deliver a bit error rate of

$$\text{BER} = Q\left(\sqrt{\frac{E_b}{N_0}}\right), \tag{6.3}$$

where Q is the error function according to

$$Q = \frac{1}{\sqrt{2\pi}} \int_x^\infty e^{-\frac{t^2}{2}} \, dt. \tag{6.4}$$

Degradations are caused by the non-ideal system components, including the channel. The graphs of 140 and 280 Mbit/s are identical since the pulse repetition time is large enough compared to the spreading properties of the system components. A data rate of 386 Mbit/s corresponds to the case where the pulse repetition time is only the sum of the pulse duration and the PPM offset T_{PPM} without any guard time. This leads to an increased intersymbol interference and hence worsens the performance. The bit error rates can also be expressed versus the SNR. The relationship between the SNR (S/N) and E_b/N_0 is:

$$\frac{E_b}{N_0} = \frac{S}{N} \cdot \frac{B}{R}, \tag{6.5}$$

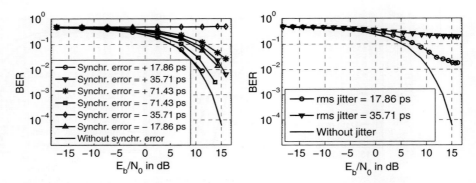

Figure 6.17 Left: BER vs E_b/N_0 with synchronization error; right: BER vs E_b/N_0 with jitter. ©2010 KIT Scientific Publishing; reprinted with permission from [168].

where B is the bandwidth (see [155]). The resulting behavior of the BER versus SNR for various data rates is presented in Fig. 6.16. A changed data rate R results in a curve shifted by the second expression of (6.3). However, to compare the bit error rates for different system configurations, the performance is usually only plotted versus E_b/N_0 since the expression E_b/N_0 is implicitly normalized to the bandwidth.

Influence of synchronization accuracy

As already mentioned, a coherent demodulation requires an accurate synchronization between the received signal and the template signal in the TD. The following section investigates what happens if the synchronization is non-ideal. The synchronization is assumed to be characterized by a constant timing error t_{synch}. This error can be positive or negative, where a positive error means that the template signal is positively delayed with respect to the received signal. Since the system simulation is based on discrete time steps, possible synchronization errors are multiples of the time step $T_0 = 17.86$ ps. The investigation shown in Fig. 6.17 (left) compares the bit error rate versus E_b/N_0 for different values of t_{synch}. The values investigated are typical in UWB systems. For example, [33] investigates a value of 50 ps. It can be seen that a synchronization error can cause a severe degradation compared to the case of an ideal synchronization. Furthermore, a positive synchronization error does not show the same performance as a negative synchronization error of the same value. This can be explained by the asymmetry of the pulse shape, which also forces the auto-correlation to be asymmetric. As a consequence, the cross-correlation between the receive and the template pulses is also asymmetric. Apart from the asymmetry, a general relationship between the spectral coverage of a signal and its auto-correlation properties must be considered. A broad spectrum usually causes the auto-correlation to be small [88], which rapidly decreases the performance in the case of a coherent demodulation characterized by a timing error. Therefore, a trade-off between spectral usage and the auto-correlation properties of the pulse shape may be the solution in a physical device. First results with an optimal jitter-robust pulse shape design are shown in [88]. However, as long as timing errors are small

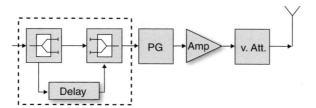

Figure 6.18 Generic UWB transmitter module for CR and ACR transceiver types. Elements in the dashed box are used for the TR system. ©2012 EuMA; reprinted with permission from [197].

enough compared to auto-correlation properties of the pulse shape, the optimization of the pulse shape in terms of power according to Fig. 6.2 makes sense.

Influence of jitter

Since jitter also introduces timing errors, the performance also degrades in the presence of jitter. Fig. 6.17 (right) shows the bit error rates for different values of the jitter root mean square (rms) standard deviation σ_{jitter}. The chosen jitter values cover the range 0–35.71 ps. They are typical for ultra-wideband applications. For example, [118] mentions typical values from 15 to 150 ps and [96] uses 20 ps rms jitter. The curves in Fig. 6.17 (right) show a saturation behavior. This principal behavior is also observed in [184], where – in contrast to the given contribution – the influence of jitter in the case of an idealized system is investigated. Furthermore, a severe degradation can be observed for $\sigma_{\text{jitter}} = 35.71$ ps since in that case the probability for the critical timing offset of -35.71 ps (see previous section) is maximized.

6.1.5   Practical transceiver implementation

In Chapter 2 the UWB transceiver architectures were described from the telecommunication engineering point of view. In this section, the transmitter and receiver concepts are considered from the RF system perspective. Based on simulations and measurements, a comparison of the three most common receiver structures that can be found in literature is performed. To enable a fair comparison, all receivers use the same components. The evaluation criterion is the signal-to-noise ratio in the processed signal using the same distances r_{TxRx} between transmitter and receiver.

Transceiver architectures

The impulse-based UWB systems are characterized by a very simple transmitter stage, where the carrierless signals are used. A short pulse of several hundred picoseconds' length is generated by a dedicated pulser circuit (pulse generator, PG), which is triggered by a baseband signal. The pulse can either be amplified (Amp.) or sent directly to the antenna. This setup is shown in Fig. 6.18.

The pulse shape generated by the PG [152] is very similar to the 5th derivative of the Gaussian pulse (PG5), with the full width at half maximum (FWHM) of 250 ps. The time and frequency domain characteristics of the PG5 are given in Fig. 6.19.

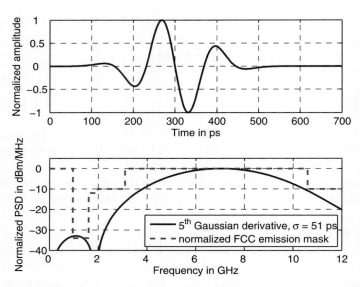

Figure 6.19 PG5 in the time and frequency domains. ©2009 IHE; reprinted with permission from [205].

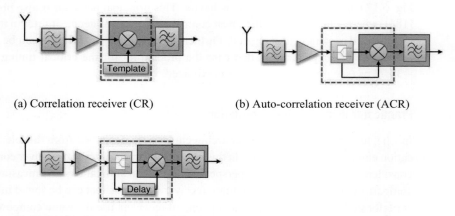

(a) Correlation receiver (CR)   (b) Auto-correlation receiver (ACR)

(c) Transmitted-reference receiver (TR)

Figure 6.20 Three investigated receiver types. The dashed boxes mark the crucial architectural differences. ©2012 EuMA; reprinted with permission from [197].

Receiver structures

In the following a brief overview of receiver structures is given, as they will be encountered in the literature [142]. The receivers can be classified based on the way the matched filter correlation is implemented: coherent or incoherent, and which pulse template is used [197].

Correlation receiver (CR) This receiver architecture (cf. Fig. 6.20(a)) is based on an analog coherent correlation of the received signal with a locally generated template. In

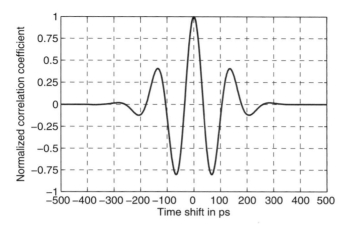

Figure 6.21 Autocorrelation function of the 5th order derivative of the Gaussian pulse.

theory, it is the most effective reception method in the AWGN channel [120]. The noisy impulse multiplied with the clean template will generate an output that is much larger than cross-correlation with noise alone. Due to the very short duration of the UWB pulses, the requirements on the timing precision are very high. The auto-correlation function of the signal presented in Fig. 6.19 is shown in Fig. 6.21. It indicates that in order to remain in the region of at least 80% of the maximum, a tolerance of only 30 ps is allowed. This results in very tight requirements for the practical system [197].

Auto-correlation receiver (ACR) An alternative, where no synchronization is required is the ACR, which is based on energy detection. The received signal is split and squared (Fig. 6.20(b)). The main drawback of this solution is the fact that the noise is correlated with itself and causes a rise of the noise floor. To deal with this, a hybrid system (TR) can be used [66].

Transmitted reference (TR) In this configuration two pulses are transmitted at a prede-fined distance, which is achieved by triggering the PG as in Fig. 6.18. The signal in the receiver is divided and one of the paths is delayed by the separation time of transmitted pulses (Fig. 6.20(c)) [32]. This ensures that the delayed copy of the first pulse (pilot) is correlated with the second pulse. The two other pulses, resulting from splitting, are multiplied with noise. The main advantage of this architecture is the fact that the time-shifted thermal noise is not correlated with itself any more. The cross-correlation of the signal and noise can however raise the noise floor as well.

Flashing receiver (FR) An even lower complexity receiver is the flashing UWB receiver [164]. This receiver mainly consists of two comparators that continuously search for positive or negative peaks (pulses) in the incoming signal. This receiver type however does not have a purely analog architecture, because the comparators (1bit ADCs) are placed directly after the LNA. The decision regarding pulse detection is made in the digital domain, and is based on a certain pattern, therefore it will not be considered in this comparison.

Quadrature analog correlation receiver (QACR) This is a type of carrier recovery-based receiver structure. The received pulse is split in two branches and down-converted to baseband with I and Q components that are then A/D converted [21, 172]. This scheme is not UWB-specific, where pulse detection is performed after the ADC (after combining the I and Q components), which makes the direct comparison with other structures impossible.

Based on these short characterizations, the first three receiver types will be compared. Due to the non-ideal hardware it is barely possible to assess which system has the best performance purely by mathematical considerations. The first three transceivers (CR, ACR and TR) were built from the same components [198] and their performance was evaluated.

Comparison criteria

In order to extract the useful information from the signal that propagated through the channel at the receiver, the ratio between the signal and noise power has to be sufficiently high. If the SNR is low, the BER increases and the transmission quality degrades. In general the receivers cope with the input SNR differently, which results in different SNRs in the processed signals. Because of this fact, the SNR in the processed signal is the most convenient way to compare receivers.

The SNR is defined as the ratio of the average power of the signal and average noise power as follows:

$$\text{SNR}_{\text{dB}} = 10 \cdot \log \left(\frac{P_{\text{signal}}}{P_{\text{noise}}} \right) = 10 \cdot \log \left(\frac{U^2_{\text{eff,signal}}}{U^2_{\text{eff,noise}}} \right). \tag{6.6}$$

The signal power is calculated from the signal's voltage $u_{\text{signal}}(t)$ and the impedance Z as:

$$P_{\text{signal}} = \frac{1}{Z \cdot T_{\text{n}}} \int_0^{T_{\text{n}}} u^2_{\text{signal}}(t) dt, \tag{6.7}$$

while the noise power results from the variance σ^2 of the noise:

$$P_{\text{noise}} = \frac{\sigma^2}{Z} = \frac{1}{Z \cdot T_{\text{n}}} \int_0^{T_{\text{n}}} n^2(t) dt, \tag{6.8}$$

with T_{n} being the duration time of the integration window. Inserting (6.8) into (6.6) results in:

$$\text{SNR}_{\text{dB}} = 10 \cdot \log \frac{P_{\text{signal}}}{P_{\text{noise}}} = 20 \cdot \log \frac{U_{\text{eff,signal}}}{\sigma}. \tag{6.9}$$

However (6.9) cannot be used for SNR in impulse-based systems without integrators in the receiving chain, because it considers the average signal and noise amplitudes.

The average for an unchanged integration window T_n depends on the signal period (or pulse repetition frequency, PRF). Because of this, the SNR varies with the changing duty

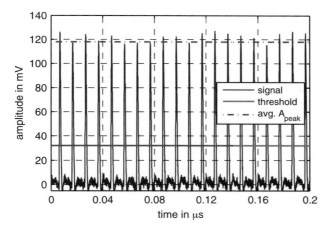

Figure 6.22 Output of the CR during optimal alignment of the received signal and the template pulse. The horizontal dashed line marks the average peak amplitude of the correlation result.

cycle of the signal. In the system where no integration is performed over several periods (the case in this investigation) the changing PRF must not influence the SNR. For this reason, another definition has to be introduced which involves only the instantaneous power (or amplitude) of the pulse, the signal-to-threshold ratio (STR):

$$\text{STR}_{\text{dB}} = 20 \cdot \log \left(\frac{A_{\text{peak}}}{3 \cdot \sigma + A_{xt}} \right), \tag{6.10}$$

where A_{peak} is the average level of the peak values of the received pulses, A_{xt} is the average peak level of the noise or cross-talk signal (e.g. from correlator template port to output) and σ the standard deviation of the noise signal. Using this definition, the calculated noise level can be directly adopted as the threshold level for the signal detection; hence the name signal-to-threshold ratio.

This parameter offers a basis for a fair comparison between different receiver architectures. The three receivers are tested at the same distance from the transmitter, and the STR gives explicit information about the transceiver performance.

In the following the estimation of STR is explained based on Figs. 6.22 and 6.23. The received signal after the correlator was recorded with a 12 GHz bandwidth oscilloscope at 40 GS/s. In the first signal the received pulses are present (for CR the optimal synchronization is needed). The second signal contains only noise (and in the case of the CR, the cross-talk of the template pulse). For statistical reasons, at every distance and for every receiver type, 300 pulses were recorded. This enabled finding the average amplitude of the received pulse and the statistical parameters of the noise signal. Substituting those terms into (6.10) results in an STR. An STR of 0 dB describes a case where the received impulse is at the threshold level. Below this point no reception is possible.

Simulation and measurement

The models of all transceivers were implemented in Agilent's ADS [15]. The models of the bandpass filter, LNA, power divider (in ACR and TR cases) and the PG

Table 6.1 Change in amplitude V_{pp} at Tx and equivalent increase in range R.

V_{pp} mV	D_{set} dB	L dB	R m
900	0	58.7	3
600	3.6	62.3	4.54
435	6.6	65.3	6.41
293	9.6	68.3	9.06
235	11.6	70.3	11.41
186	13.6	72.3	14.36
150	15.6	74.3	18.08

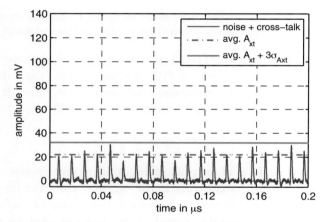

Figure 6.23 Output of the CR where no pulses were received (noise only). The cross-talk of the template pulse and its cross-correlation with noise generate prominent peaks.

were implemented, based on measurement data [198]. The correlator was modeled as a simple multiplier and the antennas as a gain block (one angular direction used in the measurement, where the value is obtained from measurements in an anechoic chamber).

The tests were performed in a laboratory. The receivers, built with the same hardware components, were placed 3 m from the transmitter and tested consecutively. To investigate the STR performance over a Tx–Rx separation, a variable attenuator was used at the Tx site (Fig. 6.18). Seven different attenuation settings were applied. The attenuation added to the free-space attenuation results in a theoretical increase of the Tx–Rx distance (see (2.12)). The attenuation settings (D_{set}), the resulting Tx signal amplitude V_{pp}, equivalent path loss L and resulting equivalent distance R are listed in Table 6.1.

The simulation and measurement results for all investigated receiver types are presented in Figs. 6.24 and 6.25, respectively. The PRF during the measurement and simulation was set to 10 MHz. The principal shapes of the curves in the simulation and measurement coincide well. The largest differences occur at small distances and

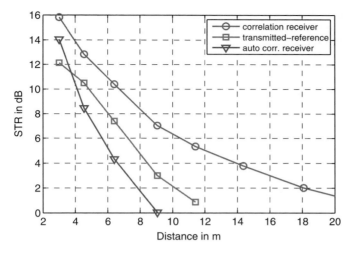

Figure 6.24 STR as a function of Tx–Rx separation: simulation results. ©2012 EuMA; reprinted with permission from [197].

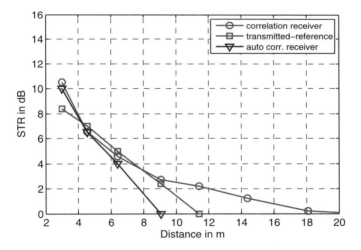

Figure 6.25 STR as a function of Tx–Rx separation: measurement results. ©2012 EuMA; reprinted with permission from [197].

relatively large signal amplitudes. The reason for this is the simplified modeling of the correlator (no cross-talk included). The differences at distances greater than 6 m for the ACR and greater than 8 m for the CR and TR are less than 1.5 dB.

At distances of 4–6 m the TR seems to be better than the CR. The reason for this is the previously mentioned small margin for the allowed synchronization error of the CR. It is highly probable that the reason for this behavior is the thermal drift of the delay element. Due to those slight changes in time delay the auto-correlation function does not reach its theoretical maximum and the CR performance degrades.

The investigation above shows that the theoretical advantage of the correlation receiver is, in general, not prominent. The CR can only outperform the other two receivers at larger distances. An advantage however is only present if the ideal synchronization is assured. Even a time misalignment as small as 20 ps between the received and template pulse can significantly degrade the performance. The auto-correlation receiver is a good choice if the architecture is to be kept simple and the signal-to-noise ratio is high (either for very short ranges or high transmission power). The transmitted-reference receiver offers the best trade-off between complexity and achievable transmission range. This advantage comes at a price due to a 50% lower transmit power efficiency (twice as many pulses required for reception as in the CR and ACR).

This shows that in practice there is no simple classification of the UWB receivers with regard to their performance. The intercept points of the STR curves indicate that there is no clear best solution and a compromise has to be made. It is also worth noting that precise modeling of hardware is essential for accurate performance prediction.

6.2   UWB localization

Localizing objects inside closed facilities has become a common requirement. An example could be the tracking of persons in offices or the positioning of automated guided vehicles in industrial environments. Being able to perform this in a precise way can open up a range of new possibilities for more advanced applications. One of the technologies that can be considered in this context is UWB. Due to its large signal bandwidth and short duration in the time domain, UWB seems to be the perfect candidate for precise distance and localization measurements. In this section several aspects of UWB positioning will be handled. For localization of a mobile unit (MU) several base stations (BS) at well known positions are required. Depending on the system concept, the MU can be the transmitter (Tx) or receiver (Rx).

6.2.1   Positioning techniques for UWB systems

The position of an object can be determined by using one or more of the following measured signal parameters [57]:

- received signal strength (RSS)
- angle of arrival (AoA)
- absolute time of arrival (ToA)
- relative time difference of arrival (TDoA).

Based on these measured signal parameters and the known coordinates of the BSs, the position of the autonomous object (here: MU) can be estimated. Besides algorithms which are based on mathematical evaluation of the parameters given above (e.g. traditional triangulation; positioning algorithms), scene analysis or proximity can be utilized to allow for a simple, but only coarse, position estimation. Each of these algorithms has

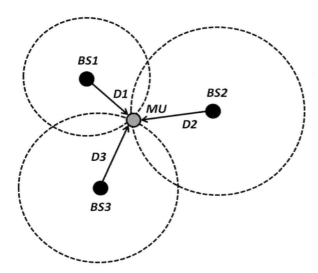

Figure 6.26 Positioning based on ToA measurements.

unique advantages and disadvantages. Therefore, using more than one type of algorithm at the same time will yield better system performance and accuracy [122].

Triangulation

Triangulation uses the geometric properties of triangles to estimate the object location at the intersection of position lines, obtained from a set of related parameters at a number of base stations. It has two derivations: *lateration*, using distance measurements, and *angulation*, using primarily angle or bearing measurements [64].

Lateration

Lateration computes the position of an MU by measuring the distance to multiple BSs. The three major methods of estimating this distance are summarized below.

1. ToA: the distance from the target node (here: MU) to the BS is directly proportional to the propagation time. In order to enable 2D positioning, ToA measurements must be calculated with respect to signals from at least three base stations as shown in Fig. 6.26. To prevent ambiguity in ToA estimates, all transmitters and receivers have to be precisely synchronized and carefully distributed at the scene.

   The performance of the ToA-based systems is sensitive to the bandwidth and to the occurrence of NLOS conditions; however, the most significant issue in the case of an autonomous MU is the accurate synchronization between the MU and the BS [56, 147]. A scenario in which ToA can be well utilized is in two-way ranging [38]. In this case the mobile unit measures the round trip travel times of the impulse to the BS and back and, based on this, calculates the relative distance [48].

2. TDoA: conventionally, ToA-based range measurements require synchronization between the target (MU) and the reference nodes (BSs). TDoA measurements can however be obtained even in the absence of synchronization between the MU and

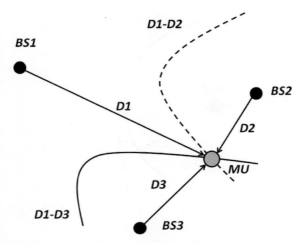

Figure 6.27 Positioning based on TDoA measurements.

the base stations. The only requirement is the synchronization among the BSs themselves [56, 97]. This allows measuring the difference between the arrival times of two signals traveling between the mobile unit and the two base stations. A hyperbola can be calculated from this time difference, measured between two BSs. The hyperbola is a set of points indicating a constant distance (or time) difference between two BSs, and it specifies all possible locations of the MU. Using at least three base stations, a 2D MU location can be estimated from the intersections of at least two hyperboloids. This is shown in Fig. 6.27. For an unambiguous 3D localization, the intersection of three hyperboloids (four BSs) is required.

3. RSS: the power of a signal transmitted between two nodes is a signal parameter that contains information related to the distance between those nodes. The RSS information can be used in two ways. The first approach is to map the power of the received signal to the distance traveled by the signal from the BS to the MU. Knowing the transmit power of the BS and the frequency, the distance can be calculated by rewriting (2.12). By measuring the RSS from at least three transmitters, the MU can be located by triangulation. In indoor scenarios however, the free-space propagation cannot be assumed and therefore path loss based RSS is not applicable.

The second approach is to use the RSS in a fingerprinting scheme [22, 137] and gather the information about the application scenario in a measurement campaign. A database containing the information about RSS from all BSs at different positions in the scenario is created, which can then serve as a look-up table. The measured RSS values are then compared with this database to obtain the object's location with greater accuracy.

Angulation

Angle of arrival (AoA) measurements provide information about the direction of an incoming signal, i.e. the angle between the two nodes. As shown in Fig. 6.28, AoA

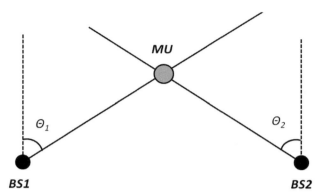

Figure 6.28 Positioning based on AoA measurements.

methods use at least two known base stations (BS1 and BS2), and two measured angles θ_1, θ_2 to compute the 2D location of the target.

The advantages of AoA are that a position estimate may be determined with only two measurement units for 2D positioning. For 3D localization at least three receiving units (not placed in-line) need to be integrated into each BS. The main advantage is the lack of requirement for time synchronization [134].

Scene analysis and proximity

The scene analysis method uses the available information, such as training data (finger-prints), and estimates the position of the target node by matching current measurements with the closest a priori location from fingerprints. To perform this, pattern-matching algorithms such as k-nearest neighbor (k-NN), support vector regression (SVR), and neural networks can be used. RSS-based localization is commonly used e.g. in WLAN.

There are two stages for local fingerprinting: *offline* and *online*. In the offline phase, the location-related data (e.g. signal strength) is measured at multiple positions in the scenario, along with the true position (e.g. the distance to the surrounding walls). During the online position calculation phase, the location-related data of an MU object is measured and compared with the pre-measured data collected in the offline phase. The closest similar entry in the database is searched to derive the location estimation [56, 97, 137].

The proximity location sensing technique examines the location of a target object with respect to a known position or area. This technique requires a number of detectors to be placed at known positions. When a tracked target is detected by a detector, the position of the target is considered to be in the proximity area marked by the detector. This only gives a rough estimate and is highly dependent on the number of pre-deployed detectors.

Conclusion

The RSS method, and connected with it a scene analysis, is rather unsuited for use in the UWB indoor localization system. The reason for this is the multipath behavior of the channel producing multiple echoes (problems with distinguishing the main path). Additionally, to assess the signal strength precisely, either an integration over several pulses would be necessary (which is problematic due to a very small duty cycle of UWB signals and leakage of integrator circuits), or a very precise A/D conversion would have to be employed. ADCs with a high number of resolution bits and a high input bandwidth (several GHz required) are barely available and power consuming.

The ToA method is not suited for UWB either, because of enormous synchronization effort (not realizable in practice in an efficient way) [199]. The symmetrical double-sided two-way ranging (or simply two-way ranging) is the specific version of the ToA system where the double time of flight between BS, MU and back is measured [38]. In this method the synchronization requirements are significantly relaxed.

The methods that are well suited for UWB localization are the TDoA and AoA, as they are already in the simplest system configuration. In a TDoA system, only the distributed reference nodes have to be synchronized (e.g. by cable), which is straightforward to realize. In the AoA case, every reference node needs to consist of multiple receivers arranged in an array; in this case only, they would need to be synchronized with each other. The localization system, combining both TDoA and AoA, yields even better results than each of the systems stand-alone [183]. In [170] an example of a commercial system employing both of these techniques is shown.

6.2.2   Steps in the UWB localization system design

In this section the most important design steps of UWB localization systems are explained in detail: starting from the choice of the algorithms, through characterization of the scenario and accuracy predictions, system setup, analysis of potential measurement errors, and finally to localization examples. Notes on how to extend the localization into a tracking system then follow. All considerations use an example of a TDoA localization system.

Relating the time differences to the MU position

To calculate the MU position from the measurement data, the error function between the true and estimated positions has to be minimized. It is however not possible to investigate the value of this function for every input argument. Because of this, a number of dedicated algorithms for solving such nonlinear problems were developed. They all follow the same pattern: first a rough estimation of the solution is done, which can then be interpreted as a start point in the error landscape. From this point, a descent direction in the error landscape is calculated in such a way that the reduction of the error function value is highly probable. Subsequently, a first step with a certain step width is done from the starting point in the descent direction. A new point is reached and serves as a new starting point to apply the same procedure, until a stop criterion is met. This can either be the change in the error function or the change in the calculated position. If this value is small enough to assume the stationary condition, a global or local minimum is reached.

The following algorithms are most suited for finding the positioning solution in a TDoA system:

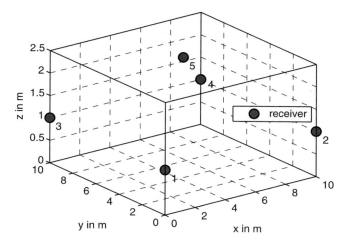

Figure 6.29 Virtual room for the evaluation of TDoA solution algorithms with five base stations (BS) and one mobile unit (MU) as receiver.

- Gauss–Newton (GN) with the *quadcubic*-line search procedure (qLSP)
- Levenberg–Marquardt (LM)
- trust region reflective algorithm (TRR)
- interior point (IP) and the modified Bancroft algorithm (BA) [202].

The comparison criteria are the following:

- the mean computation time, calculated as the average of computation times for a set of positions
- the accuracy of the solution, where the quality factor is the average 3D positioning error, calculated according to:

$$\text{mean 3D error} = \frac{1}{M} \sum_{k=1}^{M} \left\| \vec{r}_{\text{MU}_k} - \hat{\vec{r}}_{\text{MU}_k} \right\|. \tag{6.11}$$

The k in this equation denotes the number of positioning solutions, $\vec{r}_{\text{MU}_k}$ the true position and $\hat{\vec{r}}_{\text{MU}_k}$ the estimated position of the MU.

Because of the fact that some of the evaluated algorithms do not have any additional constraints (e.g. volume in which the feasible solution should remain), large positioning errors can occur and afflict the average value. For this reason, the median value is also given.

For the evaluation, a virtual room with dimensions $10\,\text{m} \times 10\,\text{m} \times 2.5\,\text{m}$ with five BSs is used. Four of the receivers are placed at a height of 1 m in the corners; the fifth one is placed in the center of the ceiling. This constellation is depicted in Fig. 6.29.

The TDoA data, used for the evaluation, is obtained by subtracting the pseudo-ranges between the BSs and the MU, including the artificially added error. The measurement noise is modeled as normally distributed, with the standard deviation $\sigma_t = 333$ ps (corresponding to $\sigma_d = 10$ cm). The assumption of homoscedasticity of the time error

Table 6.2 Average 3D error and computation times of various algorithms [204].

Algorithm	Average computation time ms	3D positioning error mean/median m
Modified Bancroft (BA)	0.580	0.406/0.312
Gauss–Newton (GN)	31.85	0.386/0.276
Levenberg–Marquardt (LM)	15.41	0.386/0.276
Trust region reflective (TRR)	40.96	0.386/0.276
Interior point (IP)	98.10	0.311/0.251

on every BS is made only for the sake of algorithm testing and does not generally apply in practice. The MU positions are picked within the scenario boundaries by a random function with equal distribution. The starting point for the iterative algorithms is chosen by the Bancroft algorithm. The results of $M = 1000$ positions calculations are presented in Table 6.2. As a constraint for the interior point and Bancroft algorithms, it is implied that the solution has to be within the room.

The first impression is that the GN algorithm delivers good results. However if the starting point were picked in a random manner and not by BA, convergence problems occur. From the algorithms that are left, the BA has the shortest computation time and its accuracy is lower (i.e. larger error) than in the case of iterative algorithms. The LM and TRR are in a similar accuracy range, however the LM requires less computation time. The IP delivers the most precise results, however this advantage is achieved at the expense of the longest computation time of all of the evaluated algorithms [204]. Based on these results, the best combination seems to be the starting point determination based on the BA and adjacent final calculation with LM [102]. In the case where the additional conditions regarding the geometry need to be accounted for, the IP is a good choice.

Scenario characterization and error prediction
The crucial aspect in localization systems is the optimal choice of reference node positions. In [153, 186] it was shown that the geometrical constellation of the MU/BS has a deciding influence on the positioning precision. The parameter that describes the quality of the geometrical setup is the geometrical dilution of precision (GDOP); a factor which states how strongly an imperfect time measurement will influence the positioning precision. In the TDoA case, when knowing the standard deviation of time measurement (σ_{time}) and a GDOP value, the positioning precision can be predicted as follows [200]:

$$\sigma_{3Dpos} = DOP \cdot \sigma_{time} \cdot c_0, \qquad (6.12)$$

where DOP represents the dilution of precision. According to the above, the more uniform the DOP distribution in the scenario, the less variation of positioning error will be observed.

Knowing this, the constellation of the reference nodes in the scenario will be discussed in more detail. The example distribution of the DOP values, split on HDOP (horizontal) and VDOP (vertical) for different reference node positions in a 20 m × 20 m × 2.5 m

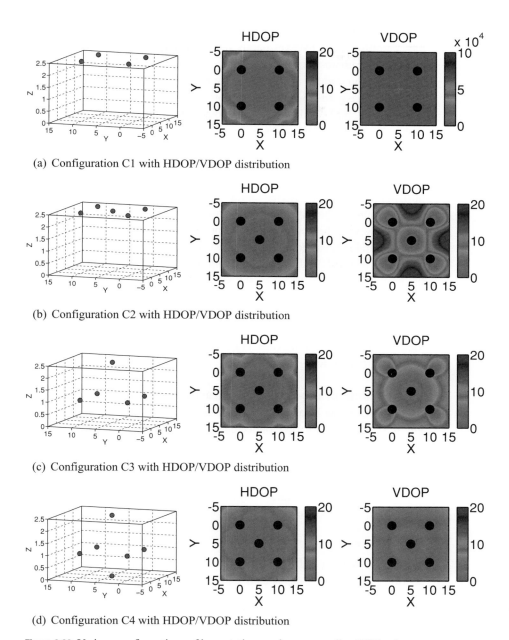

(a) Configuration C1 with HDOP/VDOP distribution

(b) Configuration C2 with HDOP/VDOP distribution

(c) Configuration C3 with HDOP/VDOP distribution

(d) Configuration C4 with HDOP/VDOP distribution

Figure 6.30 Various configurations of base stations and corresponding DOP values.

room are shown in Fig. 6.30. The configuration in Fig. 6.30(a) shows four receivers placed at a height of 2.5 m. HDOP inside the constellation is in the range $[1, \ldots, 5]$ and increases up to 10 outside of it. The VDOP distribution however is not uniform and reaches very high values in the middle of the room where the vertical movement of the target would not cause any change in the measured time differences. This situation can be avoided when an additional receiver is employed, shown in Fig. 6.30(b). In configuration C2 the centrally placed receiver barely influences the HDOP, however it does ensure that the HDOP remains in the range [2, ..., 10]. The VDOPs can be further improved by shifting the middle station out of the plane of all the other receivers or lowering the others down, e.g. to 1.5 m. This situation is depicted in Fig. 6.30(c). The VDOP became better [2, ..., 3.5], while the HDOP did not undergo any significant change.

Another measure for improving the resolution in the vertical direction is to position an additional station underneath all the present receivers. This causes longer measured time differences between two neighboring transmitter positions (Fig. 6.30(d)), which leads to a VDOP in the range [1.5, ..., 2.5]. It is worth noticing that, in all cases, the system horizontal accuracy decreases rapidly as soon as the transmitter is outside the receiver constellation. In theory an almost infinite number of different configurations could be tested, however when considering the practical aspects of base station placement in an average indoor scenario, additional constraints apply. Hence the line-ups such as C4 should be avoided. Although the distribution of DOP values is most homogeneous on the ground or slightly above it, any station placed there will most likely not be visible to an MU due to shading effects caused by inside facilities. It has to be mentioned that a larger number of base stations would give a rise to more uniform distribution of DOPs and better performance in terms of shadowing; however, the cost of practical implementation would increase. For these reasons it is advisable to use distribution C3, or similar, in practice.

System configuration and time measurement

An example of the localization system layout based on a TDoA technique is shown in Fig. 6.31. The BSs are interconnected to ensure synchronization and an autonomic MU equipped with a UWB tag transmits the impulses. The signals propagate through the scenario on physically different paths (channels) to reach the BSs. The received signals from all the BSs are forwarded to the time measurement unit through the synchronization network and undergo certain delays ($T_{\text{stop N}}$) before they trigger the time measurement at the TDC. This can be expressed as follows:

$$T_{\text{meas N}} = T_{\text{channel N}} + T_{\text{BS}} + T_{\text{stop N}} + T_{\text{offset}}, \tag{6.13}$$

where T_{BS} stands for the BS specific delay, e.g. in the RF path (between the antenna and the ADC). T_{offset} originates from the fact that the MU is not synchronized with the BSs and the transmission time point is unknown. The first impulse received triggers the time measurement and the differences to the following impulses are calculated. Such a system requires an initial calibration to determine the T_{BS} and $T_{\text{stop N}}$. After subtracting the $T_{\text{meas N}}$ equations from one another (to obtain the time differences), T_{offset} is eliminated.

For precise time measurements, a time-to-digital converter (TDC) is used. This device measures the time elapsed between the appearance of two (or more) signals at its input ports. The time difference estimation is based on the propagation time through logic gates. The first incoming signal generates a *start event* and the following ones, the *stop events*. Details can be found in [76]. An example of such a device is the commercially

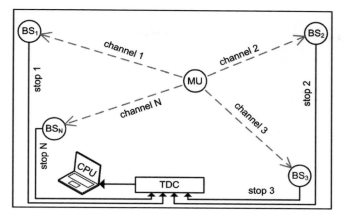

Figure 6.31 Example scenario for TDoA localization with N synchronized access points and one mobile user [204].

available ATMD-GPX produced by Acam® [2]. Depending on the operating mode, it is capable of detecting two incoming digital signals with up to 27 ps resolution.

Sources of errors in a TDoA system

An antenna is one of the most crucial devices in any wireless system. It is responsible for matching the system impedance to the free-space impedance. Real antennas however always introduce a certain signal distortion during radiation [179], which in most cases is angle dependent [127]. The parameter describing the time domain characteristic is the antenna impulse response (AIR) $h_{ant}(t)$. The shape of the AIR and the delay of its maximum can cause an additional time offset during the TDoA measurements. Because of this, in the general case the localization accuracy will be dependent on the relative angle under which the BS antenna is oriented with respect to the MU. In practice, there is no initial information about the MU orientation, hence the signals should be sent and received (to/from) all directions equally to ensure the connection to all the BSs. This implies that the MU antenna should exhibit an omnidirectional radiation pattern. On the other hand, the BS antennas will preferably have a certain directive pattern which will illuminate only a certain part of the scenario.

A possible candidate for the mobile user antenna is a planar *monopole* or *monocone* antenna, exhibiting a characteristic close to omnidirectional (in the horizontal plane). Due to the time difference approach, the influence of the MU antenna can be neglected, because the distortions can be safely assumed to be the same in each direction. For the BS, antennas with a directive radiation characteristic (e.g. Vivaldi-type) can be used. In Fig. 6.32 the distortion of the BS AIR is visualized, dependent on the angle. The time shift of the AIR's maximum during the change of angle (horizontal plane) from the main radiation direction to the perpendicular position is 260 ps.

BS antennas will preferably be placed in the corners of the room (compare Fig. 6.30) and pointing towards its center, hence the maximum transmission/reception angle should not exceed ±60°. Nevertheless, an additional time delay of up to 200 ps can be added

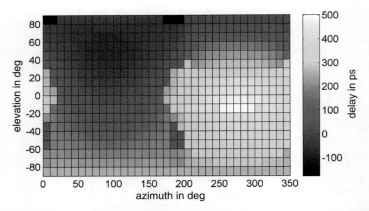

Figure 6.32 Simulation of the angle-dependent impulse response delay of the Vivaldi antenna, used as BS in [204]. The main radiation direction is (azimuth $= 90°$, elevation $= 0°$).

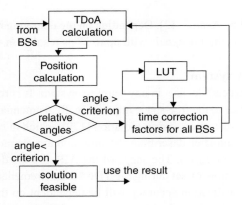

Figure 6.33 Compensation scheme for the AIR. ©2012 Hindawi; reprinted with permission from [204].

to the signal by the antenna if the received UWB pulse comes from an angular range of $\pm60°$. The explanation of the physics of this phenomenon, along with the measurement verification, can be found in [203]. This time delay corresponds to a distance of 6 cm in free space. Such an offset can have a significant impact on the accuracy of the entire system. This especially becomes an issue when attempting to reach accuracy in the lower centimeter range.

In order to eliminate the influence of the BS antenna, an iterative approach proposed in [203] can be used. The requirement for this algorithm is knowledge of the spatial orientation of the BS antennas. This can easily be assured during the system deployment phase. The flow chart of this algorithm is depicted in Fig. 6.33.

After obtaining a valid TDoA measurement, the first step is the standard calculation of the positioning solution. This leads to a first solution, which serves as a starting point for the iteration. Knowing the approximate MU location, the relative angles between the BSs' reference directions and the estimated MU position can be calculated. For those

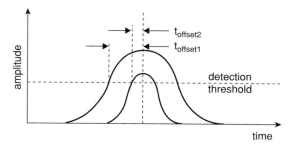

Figure 6.34 Trigger time dependency on threshold level.

angles, the time correction factors for each BS can be obtained from the look-up table (LUT). The LUT contains information about the delay of the AIR's peak, relative to the reference direction (e.g. main lobe direction), for all angles (Fig. 6.32).

After subtracting the correction factors from the original time differences, the new position can be calculated. This operation can be re-performed until a break criterion is met, e.g. if the change in the relative angle between two iterations is lower than a certain value. Another type of criterion would be the change in the calculated positioning solution. The resulting position can now be further used for applications such as tracking.

The reason why the delay correction needs to be applied is because of the choice of base station during both antenna measurement and system deployment: the feeding point (SMA-jack). A more correct and less complicated method would be to choose the phase center of the antenna (determined either by simulation or measurement) to serve as a reference. This could be done in the case of some antennas presented in this book; however, the phase center of the Vivaldi antenna, for example, is not stable over the signal's frequency range for radiation directions other than the main direction, in which case the LUT method is more reliable.

Threshold detection

In every wireless system there is an interface between the analog and digital domains, where both amplitude and time errors can appear. Depending on the digitizing device (comparator or ADC with more than 1-bit resolution) the amplitude error will have different values. Obviously the ADCs with 8 bits, or even more, are capable of transforming the signal into the digital domain with only marginal amplitude distortion. The problem with high resolution ADCs nowadays is their limited bandwidth. This, together with their cost and power consumption (e.g. pipeline ADCs consume over 1 W), results in few applications in UWB systems for the mass market. Comparators are less accurate but far cheaper, and seem to be a much better alternative for this application. These devices can achieve bandwidths close to 10 GHz and equivalent input signal rise times of 80 ps [79]. The problem that has to be addressed is the choice of threshold level. In a scenario with a large dynamic range, the trigger time dependency on the signal level will play an important role. This is depicted in Fig. 6.34. The optimal solution would be the use of a constant fraction discriminator or an adaptive threshold, offering a fixed peak-to-threshold ratio.

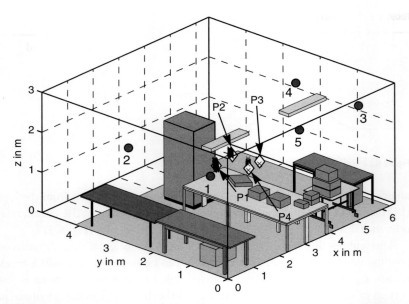

Figure 6.35 Positioning scenario for verification of the positioning system performance. BS positions are marked with numbered circles; those of MU with rhombi.

The influence of the threshold level in certain scenarios can be mitigated in a similar way to the case of the AIR. After calculating the initial positioning solution for the MU the time corrections have to be made. Knowing the exact BS coordinates and the estimated coordinates of the MU, the differences in distances between all BS–MU pairs can be extracted. Distances correspond to signal attenuation (e.g. based on free-space path loss) and this is connected with the received signal amplitude. In the LUT the estimated relative time trigger errors are saved, which are distance difference dependent. The rest of the correction procedure is same as in the case of the AIR.

The trigger time uncertainty of the comparator, caused by electronic jitter, can be modeled as a stochastic process with normal distribution. The influence of this can be minimized by performing averaging of the measured values [79]. For similar distances, e.g. if the MU is placed near to the center of the BS constellation, this effect is negligible. The reason for this is the TDoA procedure, which cancels all common time offsets.

6.2.3   Example results of time difference of arrival based UWB positioning

Based on the methods described previously in this section, a TDoA-based UWB positioning system was designed and deployed in laboratory [196]. The setup consisted of a total of five base stations, deployed in an area of 4 m × 6 m. The BS heights were chosen in such a way that the distribution of DOP values was as uniform as possible. The geometry of the scenario is shown in Fig. 6.35. The BS antennas 1 and 2 were pointing along the positive x-axis, and 3 and 4 in the negative direction. The 5th antenna was pointing along the positive y-axis, tilted slightly downwards (negative z-direction).

Table 6.3 Positioning statistics for the scenario from Fig. 6.35 (values in cm).

Position index	Corresponding DOP (no units)	Predicted error	Calc. mean error	Mean error (100-avg.)	Removed outliers	LUT
P1	5.55	11.97	10.15	6.29	3.52	2.83
P2	5.73	9.15	9.27	7.91	7.89	3.36
P3	4.74	8.22	7.38	3.01	1.80	1.35
P4	4.59	8.84	8.17	5.27	3.26	3.19
ø	–	9.55	8.74	5.63	4.11	2.68

The mobile user was placed at four different positions on the table in the middle part of the room (marked with rhombi). Before performing the TDoA measurements, the cable delays connecting the BSs with the TDC were first calibrated according to (6.13). With the calibrated system, 1000 time difference measurements at each of the four positions were taken. They were repeated for different amount of time difference averaging.

The measurement results are summarized in Table 6.3 and the accuracy improvements by applying techniques described in this chapter are presented. The indexing of MU positions in the first column corresponds to the ones from Fig. 6.35. In the second column the DOP values at positions 1–4 are listed. By calculating the standard deviation of 1000 TDoA measurements at those positions (no averaging) and multiplying it with the DOP (according to (6.11)) the error prediction can be made (column 3). The prediction coincides very well with the mean positioning error obtained from 1000 not-averaged measurements in column 4. Due to the fact that the TDoA measurement error of the TDC is normally distributed (cf. Section 6.2.2), the localization results can be greatly improved by using averaging (column 5). By simple rejection of the values that are not plausible by means of *velocity filtering* (described in Section 6.2.4) the results can be further improved. The last step (column 7) is the application of the iterative correction algorithm, accounting for the angle-dependent signal delay in the antenna, introduced in Section 6.2.2. The accuracy improvement between the last two steps is highly dependent on the initial orientation of the BS antennas with regard to the mobile user.

6.2.4   From localization to tracking

Tracking is possible in the case where the position information, obtained from the localization algorithms, is available over a period of time – so the change in position over time can be observed. The accuracy of the tracking results can be enhanced in comparison to the pure localization of single positions, by introducing additional boundary conditions such as limited velocity of the mobile user. However, the first issue that has to be overcome in localization systems is that the algorithm that estimates the position has to be supplied with valid data (e.g. time differences). Signals coming from NLOS situations have to be detected in order for them to be excluded, as they include a time bias. One possible method is the observation of velocities, resulting from discrete differentiation of position estimates. In the following, the methods that can be applied to improve the tracking results are discussed.

Receiver autonomous integrity monitoring

The offsets, caused by time of flight delays of the pulses, are one of the major sources of error during position estimation. However, if during a certain measurement more receivers than the required minimum are available, this redundancy can be utilized for error monitoring. For this purpose different receiver autonomous integrity monitoring (RAIM) algorithms can be used.

We will now describe in more detail the range comparison method (RCM) [201]. In this method the multiple position estimations are conducted with subsets consisting of four BSs (the minimum number for 3D localization). If the total number of receivers is $N > 4$, there are in general $\binom{N}{4}$ position calculations possible. The subsets are then modified by including the initial unused receivers, one after the other, and new position estimations calculated. In the subsequent step the differences (residuals) between the positioning solutions performed with the original subset and the modified one are calculated. There are two distinct cases:

- *One erroneous receiver has been included in the initial calculation.* In this case an invalid position estimation is to be expected. As a consequence, the residuals of unused receivers are large, regardless of whether they were valid or not. In this case, it is impossible to identify the invalid receiver based on residuals.
- *The erroneous receiver has not been included in the initial calculation.* In this case the position estimation is accurate. The residuals to the unused, valid receivers remain small, whereby the residuals to the unused invalid receiver are large.

The differentiation between those both cases requires a minimum of two residuals to the unused receivers. Therefore a total of $N \geq 6$ receivers are needed. The major advantage of this method in comparison with other RAIM algorithms (e.g. the least squares residual and the parity space methods) known from satellite navigation systems, is its maintained stability under conditions where multiple base stations are afflicted with time bias. Lack of initial position calculation in the RCM for all the receivers is the reason for this phenomenon.

Velocity filtering

After performing RAIM and calculating the positioning solution, the knowledge of on-site specific parameters can be applied. Depending on the scenario in which the UWB localization system is to be used, the maximum allowed velocities for all moving objects must be well defined. If an object would require too high a velocity to move from the previously estimated position to the present one, the probability of an incorrect final position record is very high. The velocity can be derived from the calculated position $\hat{\vec{r}}_S$ in the following way:

$$|\vec{v}_k| = \frac{\left\| \hat{\vec{r}}_{S,k} - \hat{\vec{r}}_{S,k-1} \right\|}{t_k - t_{k-1}} \tag{6.14}$$

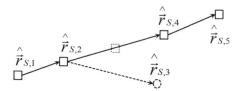

Figure 6.36 Principle of velocity filter integrity monitoring. The lengths of the arrows correspond to the estimated velocity. The velocity (dashed arrow) required to reach the calculated position $\vec{r}_{S,3}$ (dashed circle) is higher than allowed. The true position $\vec{r}_{S,3}$, fulfilling the constraints of the velocity filter, is marked with a dotted circle. ©2010 IEEE; reprinted with permission from [201].

whereby $k-1$ is the point in time of the last valid position and operator $\hat{x}$ denotes an estimated value of x. In the case where a position is found to be invalid, it is ignored in further calculations and the last valid position is used. This is illustrated in Fig. 6.36, where the estimated position $\hat{\vec{r}}_{S,3}$ is incorrect. The threshold for the subsequent velocity filtering is adapted to include double the possible distance change. In this example the velocity $|\vec{v}_4|$ has to be calculated based on positions $\hat{\vec{r}}_{S,2}$ and $\hat{\vec{r}}_{S,4}$. The classification of valid and invalid positions is based on the threshold method, whereby different maxima of allowed values can be assigned to vertical and horizontal velocity.

Hybrid tracking systems

In the case of tracking applications within buildings, the reception of radio signals from base stations can often be obstructed. By using the UWB localization/tracking system stand-alone, no positioning solution would be possible if the number of "visible" base stations were less than four (for a TDoA system). Under these conditions it is advisable to merge the UWB system with a second localization system based on another technology, e.g. inertial measurements. The known drawback of inertial navigation systems (INS) is their susceptibility to sensor noise and as a consequence only short time stability [20]. This short time stability can be used either to bridge the times when no UWB positioning is possible, or to use the measured time differences that stand-alone could not use for positioning (e.g. one or two differences), to support the INS solution (closely-coupled integration) [195].

6.3   UWB radar

Ultra-wideband offers a number of attractive features for radar and sensors, especially for short-range indoor applications. Examples include the detection and precise location of hidden objects, the detection of medical anomalies (see Section 6.5), and the characterization of reflecting targets. These are attractive applications because of the wide bandwidth and the resulting fine time resolution, unlike classical narrowband sensing such as infrared or ultrasound. Other well-known applications for ultra-wideband radar are the detection of anti-personnel mines, and breathing and heartbeat control. The hardware realization of time domain ultra-wideband radar also offers cost-attractive features.

The standard radar equation

$$P_{\text{Rx}} = \frac{P_{\text{Tx}} G_{\text{Tx}}}{\left(4\pi R^2\right)^2} \cdot \sigma \cdot A_{\text{Rx eff}} \tag{6.15}$$

is for narrowband cases; in ultra-wideband the frequency dependence of the components, especially the transmit gain G_{Tx} and the effective area $A_{\text{Rx eff}}$ of the receive antenna have to be regarded. Also the spectral power density p_{Tx} of the transmit signal is usually not constant versus frequency. These frequency dependencies require the integration of the frequency-dependent values over the operational frequency range. For a constant antenna gain, this simplifies to the integration of the frequency dependence of the effective area $A_{\text{Rx eff}}$. One possibility for this is

$$P_{\text{Rx}} = \int_{f_c-B/2}^{f_c+B/2} p_{\text{Rx}}\, df = \int_{f_c-B/2}^{f_c+B/2} p_{\text{Tx}} G_{\text{Tx}} G_{\text{Rx}} \left(\frac{\lambda}{4\pi R^2}\right)^2 df. \tag{6.16}$$

The integration here is assumed to be symmetrical with the UWB center frequency f_c.

Ultra-wideband frequency domain radar applications are, from the system point of view, similar to standard radar applications. Minor problems arise only if the group delay of the radiated signal strongly deviates from linear – but in these cases the reflected signal can easily be detected. This will not be dealt with any further here.

In the time domain, completely new problems arise for UWB radar applications because of the signal distortion by radiation, reflection and scattering of the pulses. Each of these interactions distorts the signal. The characterization of this is via the fidelity. The fidelity F of the signal compares the signal's shape to that of a reference signal.

6.3.1   UWB signal fidelity

On almost any ultra-wideband antenna, the current distribution – caused by the exciting signal – changes versus frequency. Different electric and magnetic fields are radiated versus frequency in different directions as a result. In the frequency domain this is not a severe problem, except that the radiation characteristic changes. But in the time domain it becomes severe, because the shape of the radiated impulse changes for different directions. In Fig. 6.37, this is shown for a simple monocone antenna. The shape of the radiated pulse changes with the radiating direction. This behavior is measured by the fidelity F, which relates the radiated signal to a reference signal. The fidelity is derived from the distortion d of the signal as shown in Fig. 6.38. The fidelity in direction 2 is derived from the distortion d related to the reference direction 1:

$$d = \min_{\tau} \left\{ 2 \cdot \left[1 - \int_{-\infty}^{\infty} \frac{u_2(t+\tau)}{\|u_2(t)\|_2} \cdot \frac{u_1(t)}{\|u_1(t)\|_2} dt \right] \right\} = \min_{\tau}\{2 \cdot [1-F]\} \tag{6.17}$$

The parameter d describes the distortion for the signal $u_2(t)$ compared to $u_1(t)$, taking the different travel time into account. From d results we obtain the fidelity F:

$$F = \int_{-\infty}^{\infty} \frac{u_2(t+\tau)}{\|u_2(t)\|_2} \cdot \frac{u_1(t)}{\|u_1(t)\|_2} dt \tag{6.18}$$

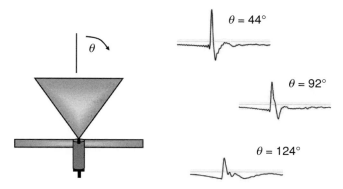

Figure 6.37 Radiated impulses by a monocone from an antenna excitation with an 88 ps Gauss pulse.

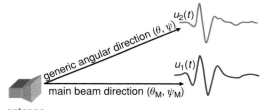

Figure 6.38 Angular deviation of radiated impulses from the main beam direction (usually the reference direction).

The maximum of the fidelity is 1, which corresponds to the reference direction. For radar applications a fidelity F of

$$0.8 < F < 1 \tag{6.19}$$

is recommended, otherwise it is difficult to correlate the receive signal with the transmit signal for time delay calculations. The fidelity can be determined in the 3D space for the designation of the radar coverage area. Targets also have a fidelity, as surface currents have similar wide bandwidths and the reflected signal is distorted similar to that from the antennas. This problem has never been properly researched.

6.3.2 UWB impulse radar measurement scheme

For radar applications, the vision is to transmit and receive a signal without distortion and interference; distortion was discussed in the last section. In order to identify the interfering signals, the block diagram of the impulse radar measurement scheme, given in Fig. 6.39, has to be analyzed.

In addition to the waves shown in the diagram, multipath behavior may occur, which can only be eliminated in the time domain by gating, or in the spatial domain by angular

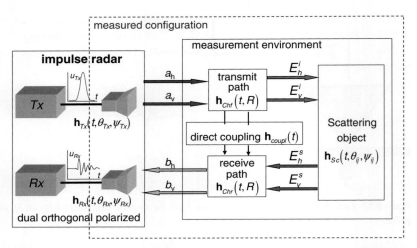

Figure 6.39 Block diagram of a polarimetric impulse radar measurement scheme. ©2009 EuMA; reprinted with permission from [126].

canceling. For all other interfering signals, calibration is the way to design a precise, reliable UWB radar. In the following a fully polarimetric calibration in the time and frequency domains for UWB radar are shown [123].

6.3.3   Polarimetric ultra-wideband radar calibration

Un-calibrated and single polarization ultra-wideband radars render only limited information as there are range, – and depending on the type of imaging – azimuth, and elevation positions of the targets to take into account. A calibrated ultra-wideband radar can improve these features by the addition of the determination of the target reflectivity. With fully polarimetric UWB radars in addition, it is possible to significantly improve the information and to characterize targets by their polarimetric signatures. Normal objects scatter the incident wave according to the target size, shape, material and composition, depending on the operating frequency in dedicated directions with specific polarizations. The determination of the polarimetric scattering allows the extraction of detailed object characteristics as a function of polarization.

The radar cross-section (RCS) of objects in a polarimetric, monostatic or quasi-monostatic ultra-wideband radar application can be expressed by the 2×2 radar cross-section matrix

$$\boldsymbol{\sigma} = \begin{bmatrix} \sigma_{hh} & \sigma_{hv} \\ \sigma_{vh} & \sigma_{vv} \end{bmatrix}. \tag{6.20}$$

By extending to the matrix of four complex quantities, the σ from (6.15) became fully polarimetric. In the above equation, the second index characterizes the polarization of the target incoming wave, the first the reflected wave polarization. The radar cross-section σ is power-related, according to (6.15).

For (polarimetric) UWB radar, depending on the tasks, time domain as well as frequency domain applications may be applied. Therefore, in the following, the relations between the RCS σ and the scattering S with the known transient response $h(t)$ and the transfer function $H(f)$ are shown. The matrix versions for the components are used here, because the fully polarimetric version will be dealt with later on. The scattering S is proportional to the field strength, while the RCS σ is proportional to power; this results in:

$$\sigma = 4 \cdot \pi \begin{bmatrix} S_{hh}^2 & S_{hv}^2 \\ S_{vh}^2 & S_{vv}^2 \end{bmatrix}. \tag{6.21}$$

From this definition, the relationship between the complex polarimetric scattering matrix S to the transfer function in the frequency domain can be derived directly by $S(f) = H(f)$:

$$\begin{bmatrix} S_{hh}(f) & S_{hv}(f) \\ S_{vh}(f) & S_{vv}(f) \end{bmatrix} = \begin{bmatrix} H_{hh}(f) & H_{hv}(f) \\ H_{vh}(f) & H_{vv}(f) \end{bmatrix}. \tag{6.22}$$

The transition from the frequency domain to the time domain is via the Fourier transformation:

$$h(t) = \begin{bmatrix} h_{hh}(t) & h_{hv}(t) \\ h_{vh}(t) & h_{vv}(t) \end{bmatrix} = \text{IFT}\{H(f)\}. \tag{6.23}$$

From this equation, the relationship back to the radar cross-section σ can be derived:

$$\sigma(t) = 4\pi \begin{bmatrix} h_{hh}^2(t) & h_{hv}^2(t) \\ h_{vh}^2(t) & h_{vv}^2(t) \end{bmatrix}. \tag{6.24}$$

The 2×2 scattering matrix S is based on two orthogonal polarizations. Here the calibration for horizontal and vertical polarization is derived, but it could be similarly derived for right- and left-hand circular polarizations. In the above equations, the RCS and scattering coefficients for co-polarization are indexed vv and hh, and the cross-polarization hv and vh, both in the frequency and time domains.

Frequency domain UWB radar channel

In order to fully understand the problem and the resulting solution, the complete frequency domain radar calibration scheme is shown in Fig. 6.40. The transmitter–receiver setup is quasi-monostatic. In order to reduce the Tx–Rx coupling, they are usually separated by a small distance compared to the target range. The signal $U_{Tx}(f)$ is transmitted via the transmit antenna $H_{Tx}(f, \theta_{Tx}, \psi_{Tx})$, to the scatterer in the forward transmit channel $H_{Chf}(f, R)$. The object scatters the signal $H_{Sc}(f, \theta_{Sc}, \psi_{Sc})$ and returns it in the return channel $H_{Chr}(f, R)$ and via the receive antenna $H_{Rx}(f, \theta_{Rx}, \psi_{Rx})$ to the receiver $U_{Rx}(f)$. In (6.25), the complete signal propagation chain is given, including the coupling Tx–Rx $H_{coupl}(f, \theta_{Tx}, \psi_{Tx})$ and the background reflections. The coupling and the background reflections are both included in the additive coupling term as they are both independent

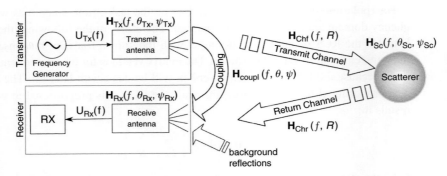

Figure 6.40 UWB frequency domain, quasi-monostatic radar system.

of the target:

$$\frac{U_{\mathrm{Rx}}(f)}{\sqrt{Z_{\mathrm{Rx}}^C}} = \boldsymbol{H}_{\mathrm{Rx}}^T(f, \theta_{\mathrm{Rx}}, \psi_{\mathrm{Rx}})$$

$$\cdot \left\{ \boldsymbol{H}_{\mathrm{coupl}}(f) + \boldsymbol{H}_{\mathrm{Chr}}(f, R) \cdot \boldsymbol{H}_{\mathrm{Sc}}(f, \theta_{\mathrm{Sc}}, \psi_{\mathrm{Sc}}) \cdot \boldsymbol{H}_{\mathrm{Chf}}(f, R) \right\} \quad (6.25)$$

$$\cdot \boldsymbol{H}_{\mathrm{Tx}}(f, \theta_{\mathrm{Tx}} \psi_{\mathrm{Tx}}) \cdot \frac{\partial}{\partial t} \frac{U_{\mathrm{Tx}}(f)}{\sqrt{Z_{\mathrm{Tx}}^C}}.$$

The matrices in the above equation are all 2×2 (see, for example, (6.22)). The measured transfer function $\boldsymbol{H}_m$ results in the dependencies $(f, \theta_{\mathrm{Tx}}, \psi_{\mathrm{Tx}})$, which are omitted for simplification:

$$\begin{bmatrix} H_m^{\mathrm{hh}} & H_m^{\mathrm{hv}} \\ H_m^{\mathrm{vh}} & H_m^{\mathrm{vv}} \end{bmatrix} = \begin{bmatrix} H_{\mathrm{Rx}}^{\mathrm{hh}} & H_{\mathrm{Rx}}^{\mathrm{hv}} \\ H_{\mathrm{Rx}}^{\mathrm{vh}} & H_{\mathrm{Rx}}^{\mathrm{vv}} \end{bmatrix}$$

$$\cdot \left\{ \begin{bmatrix} H_{\mathrm{coupl}}^{\mathrm{hh}} & H_{\mathrm{coupl}}^{\mathrm{hv}} \\ H_{\mathrm{coupl}}^{\mathrm{vh}} & H_{\mathrm{coupl}}^{\mathrm{vv}} \end{bmatrix} + \begin{bmatrix} H_{\mathrm{Ch}}^{\mathrm{hh}} & H_{\mathrm{Ch}}^{\mathrm{hv}} \\ H_{\mathrm{Ch}}^{\mathrm{vh}} & H_{\mathrm{Ch}}^{\mathrm{vv}} \end{bmatrix} \cdot \begin{bmatrix} H_{\mathrm{Sc}}^{\mathrm{hh}} & H_{\mathrm{Sc}}^{\mathrm{hv}} \\ H_{\mathrm{Sc}}^{\mathrm{vh}} & H_{\mathrm{Sc}}^{\mathrm{vv}} \end{bmatrix} \cdot \begin{bmatrix} H_{\mathrm{Ch}}^{\mathrm{hh}} & H_{\mathrm{Ch}}^{\mathrm{hv}} \\ H_{\mathrm{Ch}}^{\mathrm{vh}} & H_{\mathrm{Ch}}^{\mathrm{vv}} \end{bmatrix} \right\}$$

$$\cdot \begin{bmatrix} H_{\mathrm{Tx}}^{\mathrm{hh}} & H_{\mathrm{Tx}}^{\mathrm{hv}} \\ H_{\mathrm{Tx}}^{\mathrm{vh}} & H_{\mathrm{Tx}}^{\mathrm{vv}} \end{bmatrix}. \quad (6.26)$$

In this equation, and also in the following ones where required, the polarization indices are the top indices, and the system reference indices are the lower ones. The fully polarimetric calibration results in total in 24 calibration terms, which can normally be determined by six calibration targets. In the following, the time domain radar channel is explained, as the hardware calibration procedure is similar to the time and frequency domains.

Time domain UWB radar channel

The system components for the time domain radar channel correspond directly to the frequency domain configuration, as can be seen in Fig. 6.41. The resulting transient

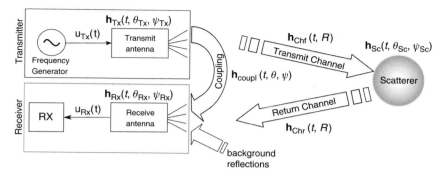

Figure 6.41 UWB time domain, quasi-monostatic radar system.

response is noted in the following equation:

$$\frac{u_{\text{Rx}}(t)}{\sqrt{Z_{\text{Rx}}^C}} = \boldsymbol{h}_{\text{Rx}}^T(t, \theta_{\text{Rx}}, \psi_{\text{Rx}})$$

$$* \left\{ \boldsymbol{h}_{\text{coupl}}(t) + \boldsymbol{h}_{\text{Chr}}(t, R) * \boldsymbol{h}_{\text{Sc}}(t, \theta_{\text{Sc}}, \psi_{\text{Sc}}) * \boldsymbol{h}_{\text{Chf}}(t, R) \right\} \quad (6.27)$$

$$* \boldsymbol{h}_{\text{Tx}}(t, \theta_{\text{Tx}}, \psi_{\text{Tx}}) * \frac{\partial}{\partial t} \frac{u_{\text{Tx}}(t)}{\sqrt{Z_{\text{Tx}}^C}}.$$

The resulting measured transient response $\boldsymbol{h}_m(t)$ can be described by the following convolution (the dependencies $(f, \theta_{\text{Tx}}, \psi_{\text{Tx}})$ are omitted for simplification):

$$\boldsymbol{h}_m = \boldsymbol{h}_{\text{Rx}}^T * \left\{ \boldsymbol{h}_{\text{coupl}} + \boldsymbol{h}_{\text{Ch}} * \boldsymbol{h}_{\text{Sc}} * \boldsymbol{h}_{\text{ch}} \right\} * \boldsymbol{h}_{\text{Tx}}, \quad (6.28)$$

or written as 2×2 matrices:

$$\begin{bmatrix} h_m^{\text{hh}} & h_m^{\text{hv}} \\ h_m^{\text{vh}} & h_m^{\text{vv}} \end{bmatrix} = \begin{bmatrix} h_{\text{Rx}}^{\text{hh}} & h_{\text{Rx}}^{\text{hv}} \\ h_{\text{Rx}}^{\text{vh}} & h_{\text{Rx}}^{\text{vv}} \end{bmatrix}$$

$$* \left\{ \begin{bmatrix} h_{\text{coupl}}^{\text{hh}} & h_{\text{coupl}}^{\text{hv}} \\ h_{\text{coupl}}^{\text{vh}} & h_{\text{coupl}}^{\text{vv}} \end{bmatrix} + \begin{bmatrix} h_{\text{Ch}}^{\text{hh}} & h_{\text{Ch}}^{\text{hv}} \\ h_{\text{Ch}}^{\text{vh}} & h_{\text{Ch}}^{\text{vv}} \end{bmatrix} * \begin{bmatrix} h_{\text{Sc}}^{\text{hh}} & h_{\text{Sc}}^{\text{hv}} \\ h_{\text{Sc}}^{\text{vh}} & h_{\text{Sc}}^{\text{vv}} \end{bmatrix} * \begin{bmatrix} h_{\text{Ch}}^{\text{hh}} & h_{\text{Ch}}^{\text{hv}} \\ h_{\text{Ch}}^{\text{vh}} & h_{\text{Ch}}^{\text{vv}} \end{bmatrix} \right\}$$

$$* \begin{bmatrix} h_{\text{Tx}}^{\text{hh}} & H_{\text{Tx}}^{\text{hv}} \\ h_{\text{Tx}}^{\text{vh}} & h_{\text{Tx}}^{\text{vv}} \end{bmatrix}. \quad (6.29)$$

In these and the following equations, the channel forward $\boldsymbol{h}_{\text{Chf}}$ and channel return $\boldsymbol{h}_{\text{Chr}}$ are assumed to be identical $\boldsymbol{h}_{\text{Ch}}$. This is not valid for bistatic measurements, but the above derivations are valid for bistatic applications.

The structure in the time domain is similar to the one in the frequency domain. The 24 unknowns on the right-hand side of the above equation have to be determined by reference targets. In practice it is not so critical, as several of these components are identical, and several others can be determined or eliminated by range gating.

6.3.4   Calibration steps

The calibration of an ultra-wideband radar system is most favorably performed in an anechoic chamber, but this is not mandatory as unwanted reflections can be eliminated by gating. The requirements for calibration targets are:

- uncritical alignment
- high radar cross-section
- RCS to be precisely calculated
- flat RCS versus frequency
- wide frequency range
- broad cross-section pattern
- targets for co- and cross-polarization.

For UWB an additional requirement is that there should be only one scattering center. This is unfortunately difficult to realize.

Antenna calibration

The first step is the bistatic calibration of the antennas. The antennas are arranged face to face and the co- and cross-polarizations are excited. Each time a single polarization is transmitted, the co- and cross-polarization on the receive side are measured; see also Section 3.1.

Empty room calibration

The above equations show the frequency and time domain contributions H_{coupl} and h_{coupl}, which generate a receive signal even without a target. Therefore in the next step, the quasi-monostatic arrangement (antennas side-by-side) is evaluated in the so-called empty room, no target present. This allows the determination of the polarimetric coupling and the background reflections:

$$\begin{bmatrix} h_m^{\text{hh}} & h_m^{\text{hv}} \\ h_m^{\text{vh}} & h_m^{\text{vv}} \end{bmatrix} = \begin{bmatrix} h_{\text{Rx}}^{\text{hh}} & h_{\text{Rx}}^{\text{hv}} \\ h_{\text{Rx}}^{\text{vh}} & h_{\text{Rx}}^{\text{vv}} \end{bmatrix}^T * \begin{bmatrix} h_{\text{coupl}}^{\text{hh}} & h_{\text{coupl}}^{\text{hv}} \\ h_{\text{coupl}}^{\text{vh}} & h_{\text{coupl}}^{\text{vv}} \end{bmatrix} * \begin{bmatrix} h_{\text{Tx}}^{\text{hh}} & h_{\text{Tx}}^{\text{hv}} \\ h_{\text{Tx}}^{\text{vh}} & h_{\text{Tx}}^{\text{vv}} \end{bmatrix}. \tag{6.30}$$

From the previous calibration, the antenna transfer functions and the transient responses are known in the frequency and time domains for Tx–Rx coupling parameter determination in the above equation.

Channel, co-polarization and polarization coupling calibration

In the next step, the channel and polarization coupling are determined. For this, targets are required that reflect only co-polarization:

$$S = \begin{bmatrix} S_{\text{hh}} & 0 \\ 0 & S_{\text{vv}} \end{bmatrix} \quad \text{and} \quad S_{\text{hh}} = S_{\text{vv}}. \tag{6.31}$$

Typical targets for this are, for example, a sphere or circular flat plate, for which the scattering is given below.

Figure 6.42 Dihedral, left: co-polarization σ_{hh} and σ_{vv} only; right: cross-polarization only σ_{hv} and σ_{vh}.

Sphere:

$$S_{hh} = S_{vv} = \sum_{n=1}^{\infty} (-j) \cdot \frac{n(n+1)}{2} \cdot (A_n - j B_n) \tag{6.32}$$

where A_n and B_n are the Mie coefficients.

Circular plate:

$$\sigma_{hh} = \sigma_{vv} = \frac{4 r^4 \pi^3}{\lambda^2} . \tag{6.33}$$

Square plate:

$$\sigma_{hh} = \sigma_{vv} = \frac{4\pi a^2 b^2}{\lambda^2} . \tag{6.34}$$

The alignment of the plates has to be orthogonal to the center of the Tx–Rx antennas; the edges of the square plate have to be vertical and horizontal.

The scattering coefficient for the target is calculated and hence the measured value is calibrated for future applications:

$$\widehat{h}_m = h_{Chr} * h_{Sc} * h_{Chf} \implies \widehat{h}_m = C * h_{Sc}. \tag{6.35}$$

One has to be aware of the delayed creeping waves of the sphere and also of the edge scattering of the flat plates, especially if misaligned. By these measurements, the matrix C can be determined and the system is ready for precise co-polarization measurement.

Cross-polarization calibration

The major problems usually occur when calibrating the cross-polarization. Here, the best results are obtained with targets that can be positioned in such a way that in the first measurement they only reflect co-polarization and in a second measurement they only reflect cross-polarization (see Fig. 6.42). One of the best targets for this is a dihedral. Positioned in co-polarization, for example vertically, the scattering of the dihedral can be precisely determined by the last step, the co-polarization calibration of the sphere or the flat plate. The dihedral RCS is:

$$\sigma_{Di} = \frac{8\pi a^2 h^2}{\lambda^2} , \tag{6.36}$$

$$\sigma_{co\text{-}pol} = |\sigma_{Di}| \cdot \begin{bmatrix} 1 & 0 \\ 0 & -1 \end{bmatrix} , \quad \sigma_{cross\text{-}pol} = |\sigma_{Di}| \cdot \begin{bmatrix} 0 & 1 \\ 1 & 0 \end{bmatrix} . \tag{6.37}$$

If the dihedral is rotated $45°$ in the next step, it reflects only cross-polarization, but the amplitudes remain identical. With these two measurements, the remaining unknowns in (6.26) and (6.29) can be determined. More detailed information on this can be found in [123].

This closes the polarimetric UWB radar calibration section. This calibration is mandatory for precise, polarimetric UWB time domain radar applications. From the system point of view, ultra-wideband frequency domain radar applications are similar to standard radar applications. Minor problems only arise if the group delay of the radiated signal strongly deviates from linear, but in this case the reflected signal can be well detected. This will not be dealt with any further here. In the time domain, completely new problems arise for UWB radar applications because of the signal distortion by radiation, reflection, and scattering of the pulses. Each of these interactions distorts the signal. The characterization of this is by the fidelity F of the signal, which relates the signal radiated in a given direction or after interaction to a reference signal, for example in main beam direction or before interaction.

6.4   UWB imaging

6.4.1   Overview of UWB imaging

In general, microwave imaging applications are important in everyday life. Compared to the alternative techniques such as computed tomography (CT) using X-ray (ionizing radiation) and magnetic resonance imaging (MRI) using strong magnetic fields, microwave imaging arises as a promising imaging technique benefiting from its non-ionization, low cost, and low system complexity. Therefore, microwave imaging is widely applied for remote sensing, scenario analysis, characterization of materials, medical diagnostics, etc. Microwave imaging systems may either be passive or active. Examples of passive systems are microwave radiometry and thermography [1]. Active microwave imaging systems include tomography, holography and radar imaging [115]. Applying the UWB technology to microwave imaging creates new possibilities in the field of object detection with radar sensors – due to the high range resolution of the signal itself. A major current research field is medical microwave imaging but, since the UWB signal features a high penetration capability, it can also be used for through-wall detection, ground penetration, and indoor navigation [8, 145, 167, 193].

UWB imaging can be divided into time and frequency domain techniques [187]. The time domain technique is based on the transmission of ultra-short pulses (in the picosecond range) with a UWB spectrum. By measuring the time delay of the scattered and received pulses, the distance of the target can be determined. Due to the huge bandwidth of the UWB signal, a very good time resolution is achievable, which corresponds to a few centimeters. The frequency domain technique is usually based on a system which sweeps over the whole UWB frequency band and measures the signal at one very narrow frequency band at a time which will achieve the highest possible system dynamic range. Since the imaging quality is strongly dependent on the measurement accuracy and

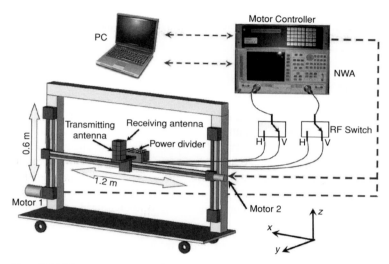

Figure 6.43 Measurement setup for indoor UWB imaging. ©2010 IEEE; reprinted with permission from [91].

system dynamic range [187], the frequency sweep approach is widely applied for UWB imaging. The obtained signals in the frequency domain can then be transformed into the time domain by an inverse fast Fourier transformation (IFFT) for the image processing.

To obtain more information about the target, a polarization-dependent measurement system is often adopted for microwave imaging. The polarization properties of targets can be utilized by an imaging radar (i.e. spaceborne SAR) to classify the target. Generally, the objects scatter electromagnetic waves in different ways for each polarization. This information can be used for the characterization of their surface, shape, dielectric properties, etc. Additionally, some objects might scatter the wave only in the cross-polarization, which makes them undetectable for monopolarized sensors (e.g. 45° rotated dihedral). These polarization properties can be used for recognition and classification of objects (see Section 6.3).

In the following sections, a UWB indoor imaging system with dual-orthogonal polarized antennas (fully polarimetric UWB indoor imaging system) is discussed in detail to demonstrate its capabilities. First, the imaging system and the image reconstruction method are introduced. The imaging system is then evaluated using different indoor scenarios, using 2D and 3D imaging.

6.4.2   Measurement setup of a fully polarimetric UWB indoor imaging system

The fully polarimetric UWB indoor imaging system is shown in Fig. 6.43. The UWB signal is radiated by the Tx antenna and the reflection is received by the Rx antenna. For more details on the antennas, see [4]. The corresponding response from the scenario is collected by a commercial vector network analyzer (VNA) in the frequency domain. To perform the fully polarimetric imaging, the RF switches are applied to obtain the radiated and received waves both in E- and H-polarization. The transfer function of

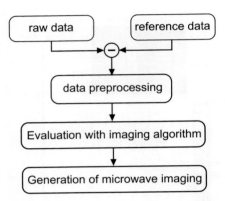

Figure 6.44 Block diagram of data processing. ©2010 IEEE; reprinted with permission from [91].

the system $S_{21}(f)$ is then measured in the frequency range 2.5–12.5 GHz in steps of 12.5 MHz four times to obtain all combinations of the Tx and Rx polarizations. By using an IFFT, the system impulse response can be then calculated from S_{21} and is used for the image reconstruction (see also Section 6.4.3).

The measurement setup allows a 2D or 3D imaging of the scenario. For 2D imaging, a scan at a certain height is performed. In this case, a 4×1 vertically expanded antenna array is used [6]. The resulting radiator has a wide beamwidth in the horizontal plane but a narrow one in the vertical plane. The wide beamwidth in the horizontal plane assures a good azimuth resolution of the processed image in the horizontal direction, whereas the narrow beamwidth in the vertical plane concentrates the radiated signal at a certain height. For the generation of a 3D image, a scan in the x–z plane is performed. In this case, a single dual-orthogonal polarized antenna is used to assure the same image properties in both horizontal and vertical planes.

The transfer functions of the antennas influence the S_{21} and consequently the system impulse response [159]. The phase centers of the radiators are constant over frequency, which assures an advantageously short impulse response of the antennas. The position of the phase centers is the same for both polarizations, which guarantees the same radiation conditions in each polarization state. The antennas show a very good polarization decoupling of more than 20 dB over the whole frequency range, which is sufficient for the fully polarimetric operation. The beamwidths in the E- and H-planes have nearly the same values, which assures similar imaging performances along the horizontal and vertical planes. The beam of the single antenna is able to successfully suppress the undesired grating lobes in the array over a wide frequency range. It can be clearly seen that the imaging application poses more stringent requirements on the antennas than typical applications like communications.

6.4.3   Image reconstruction method

A visual image of the scenarios can be obtained from the received signals by data processing. The block diagram of the data processing is shown in Fig. 6.44. The reference

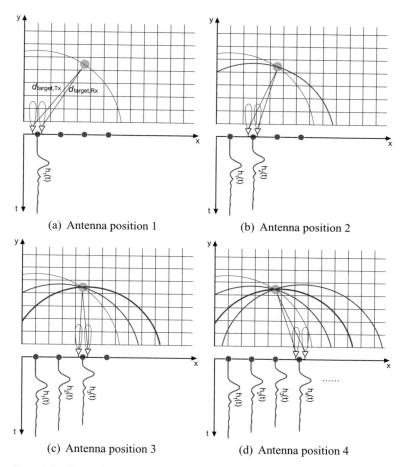

(a) Antenna position 1

(b) Antenna position 2

(c) Antenna position 3

(d) Antenna position 4

Figure 6.45 Illustration of the Kirchhoff migration.

data is obtained in the measured scenario without the object (empty room measurement). This is needed for the subtraction of the background reflections and the antenna coupling from the raw data (see Section 6.3). The influence of the microwave devices (e.g. RF switches, RF cables, etc.) is calibrated out. Zero-padding and the Hamming window in the frequency range are applied to improve the time resolution and side lobes suppression in the time domain [91]. The image algorithm is used to reconstruct the measured scenario using the preprocessed data.

One approach for image reconstruction is the Kirchhoff migration, which is an imaging algorithm in the time domain. A prerequisite to apply the Kirchhoff migration is a homogeneous medium of which the propagation velocity should be known. The image focusing technique is based on movement of the sensor and simultaneous data collection [61, 62, 145]. In the following paragraphs the algorithm is discussed in detail.

The measurement scenario is illustrated in Fig. 6.45. While the antennas are moved along a scan track in a horizontal direction, the impulse responses $h_n(t)$ for all positions are recorded (index n denotes the position of the antennas (sensors)). The time of arrival of the peak changes due to different distances between object and antenna positions.

The measurement system is a quasi-monostatic antenna system, since transmitter and receiver are placed very close together on the positioner. The target, on the other hand, must be in the far field of the antenna. Therefore, the distance between transmitter and target $d_{target,Tx}$ becomes approximately the same as the distance between receiver and target $d_{target,Rx}$ (refer to Fig. 6.45(a)). The propagation time of the electromagnetic wave can be calculated as:

$$\Delta t = \frac{d_{target,Tx} + d_{target,Rx}}{c_0} \approx \frac{2d_{target,Tx}}{c_0} \qquad (6.38)$$

where c_0 is the propagation velocity of the medium and is assumed to be the velocity of light.

For a 2D image, a grid $o(x,y)$ is created [188], which is a projection of the impulse responses at different positions on the two-dimensional illumination plane. The intensity of each grid cell is calculated with the following formula:

$$o(x,y) = \sum_{n=1}^{N} h_n(2t_n) = \sum_{n=1}^{N} h_n\left(\frac{2r_n}{c_0}\right), \qquad (6.39)$$

where

$$r_n = \sqrt{(x - x_n)^2 + (y - y_n)^2}. \qquad (6.40)$$

The (x_n, y_n) describes the antenna position and $h_n(t)$ is the corresponding system impulse response. For each antenna position the grid values $o(x,y)$ are computed and added together. The approach of the Kirchhoff migration is illustrated in Fig. 6.45 with respect to different antenna positions. The algorithm then produces image spots of high intensity, which correspond to the position and reflectivity of the target. The image contrast becomes greater when the number of impulse responses recorded at different positions is increased. In the same way, the image reconstruction method can be extended to a 3D image by modifying (6.40). With the presented approach, the images for each polarization configuration HH, HV, VH, VV can be obtained.

6.4.4   Performance of a fully polarimetric UWB imaging system

Based on the measurement setup and the image reconstruction algorithm, imaging results of different indoor scenarios will be provided in the following section. First, 2D imaging of simple shapes like metallic wires and more complex shapes are provided to investigate the performance of the imaging system in terms of detection capability, image resolution and polarization diversity. Next, the detection of hidden objects behind plasterboard is presented using 3D imaging, to demonstrate the penetration capability of UWB signals and the detection capability of a 3D imaging system.

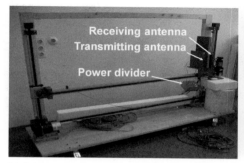 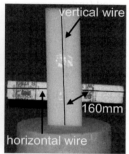

(a) Measurement setup

(b) Two thin metallic wires

Figure 6.46 Photographs of the measurement setup. ©2010 IEEE; reprinted with permission from [91].

2D imaging

Since many indoor objects consist of bars or wires oriented in different directions (such as water pipes or electric wires), they are ideal targets for a 2D UWB imaging demonstration. As explained before, a 4×1 vertically expanded antenna array is used to scan the target at a certain height as shown in Fig. 6.43. The photographs of the measurement setup and the first target are shown in Fig. 6.46. Two thin metallic wires are fixed orthogonally on a styrofoam holder. The diameter and length of the wires are 0.25 mm and 0.3 m, respectively. The horizontal wire is behind the horizontal styrofoam, and the distance between the two wires is adjusted to be 0.16 m. The vertical wire is placed at a distance of 2.25 m from the antenna. A slightly skewed fixation of the horizontal wire is applied to the antenna track, and the scan performed at the height of the horizontal wire. Since the reflection from the metallic wires is very weak due to their small diameter, the measurement is performed in the antenna anechoic chamber where the background reflection can be minimized.

The measured and processed microwave images are shown in Fig. 6.47 for each polarization channel (HH, HV, VV and VH). The image shows that the vertical and horizontal wires are detected at a distance of 2.25 and 2.40 m respectively, but in different polarization configurations. It can be seen that the vertical wire only reflects the vertically polarized EM wave, again with vertical polarization. The respective behavior is observed for the horizontal polarization where a strong reflection from the horizontally oriented wire, and only a weak reflection from the vertically oriented wire, is present. Also, the slightly skewed orientation of the horizontal wire with respect to the antenna track is detected. Hence the orientation of thin wire can be determined from the polarization information by utilizing the polarization diversity in the imaging system, which cannot be achieved using radar imaging with just one polarization.

For the 2D imaging of more complex shapes, five targets (i.e. boxes) covered with aluminum foil are aligned linearly at a fixed distance (see Fig. 6.48). Four boxes reflect the co-polarized incident wave (HH and VV) and the $45°$ rotated dihedral reflects only

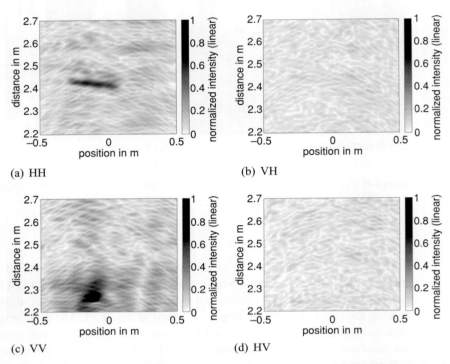

(a) HH

(b) VH

(c) VV

(d) HV

Figure 6.47 Imaging results of two orthogonally oriented thin wires. ©2010 IEEE; reprinted with permission from [91].

(a) Front view

(b) Top view

Figure 6.48 Photograph of arrangement of the targets. ©2010 IEEE; reprinted with permission from [91].

the cross-polarized components (HV and VH). From the processed imaging results in Fig. 6.49 it can be seen that all five objects are clearly resolved in the azimuth direction, which confirms the high azimuth resolution. As expected, the first four boxes are detected by HH and VV polarization, while the 45° rotated dihedral is clearly seen only in the HV and VH configuration. Moreover, it can be observed that the spots seen in the HH and VV components are located at the same distance and possess similar shapes. The HV and VH components provide similar results for the 45° rotated dihedral. This demonstrates

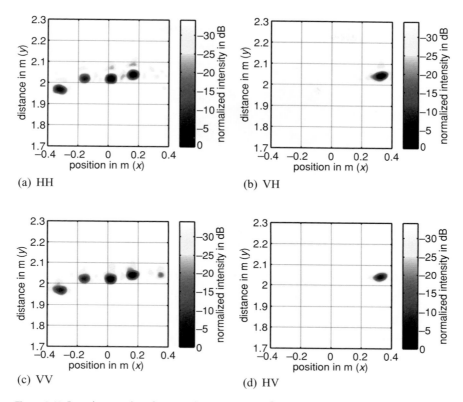

Figure 6.49 Imaging results of targets from Fig. 6.48. ©2010 IEEE; reprinted with permission from [91].

that the antenna array features the same radiation properties for both polarizations (H and V) in terms of gain, phase center and impulse response [90, 91].

3D imaging

In order to demonstrate the possibility of extending this approach to the 3D case, a new scenario is provided in Fig. 6.50. Two plasterboards with a thickness of 1 cm are placed in front of a wooden cabinet. Inside the cabinet, five objects are placed at different positions regarding all three dimensions: a metallic sphere, a metallic box, a vertically oriented dihedral, a 45° rotated dihedral and a bottle of detergent. The axes x, y, z denote the position of the antenna, the distance and the height, respectively.

The plasterboards, as well as the cabinet door, are clearly identified in HH and VV configurations (see Fig. 6.51) as expected. In addition, the locations of all the objects in the cabinet are correctly represented. The metal box, the dihedral and the detergent provide strong reflections. The sphere, on the other hand, shows a weak reflection, since the incident EM wave is scattered in different directions. The 45° rotated dihedral remains undetected in the HH and VV configuration. However, it can be clearly recognized in the cross-polarized configuration.

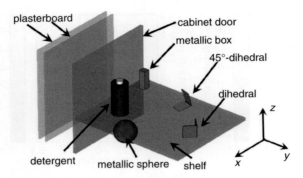

Figure 6.50 Sketch of 3D scenario: plasterboard and cabinet with objects. ©2010 IEEE; reprinted with permission from [91].

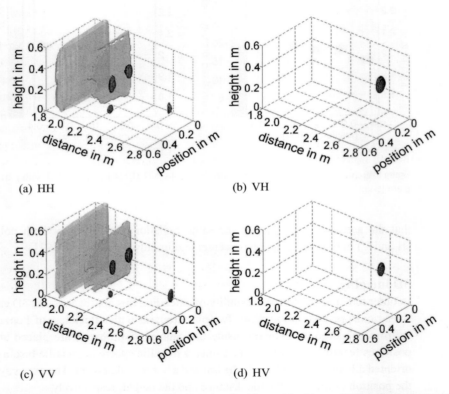

Figure 6.51 Imaging results of plasterboard and cabinet with objects. ©2010 IEEE; reprinted with permission from [91].

By UWB imaging measurements, the exact positions and properties of even small objects with different reflection behavior can be clearly identified and distinguished, especially by applying polarization diversity. Even the detection of hidden objects is possible, due to the penetration ability of UWB signals.

6.5   UWB medical applications

There is increasing interest in scientific research on the development of medical applications based on UWB techniques [167]. Compared to ultrasound, X-ray imaging and tomography, UWB signals provide an attractive alternative for imaging of the lungs or brain due to their non-ionizing effect and good penetration capability. UWB offers the potential of developing new medical devices with low cost, very high resolution or high data rate for different medical applications [18]. In this section the four most important medical UWB applications currently under investigation in research and industry are discussed (see also Section 3.5).

Monitoring of human vital functions

The electrocardiogram (ECG) is widely used for recording the electrical signals of the heart and for screening for cardiovascular disease. It typically requires placing many electrodes on the body. With a biomedical UWB radar, continuous and wireless monitoring of vital functions becomes possible. The respiration and heart rate are resolved by detecting the continuous movement of the chest and heart muscle [70]. Therefore, UWB radar can avoid the use of many electrical wires around the patient which would be beneficial for minimizing patient disruption [130]. Probably the most important application would be the monitoring of life signs such as breath frequency, heartbeats, and motion of patients in an intensive care unit (ICU). Third-degree burn victims would particularly benefit from a contactless UWB solution. An example of a UWB radar to monitor vital signs, based on a correlation receiver, is presented in [89].

Breast cancer detection

More than 45,000 women are diagnosed with breast cancer and over 131,900 women die from the disease every year in Europe [47]. The goal of breast cancer detection is to recognize the cancers at a very early stage and thus avoid the start of symptoms which tend to be larger or spread beyond the breast [175]. Thousands of women's lives could be saved through early detection. Generally, the contrast in the dielectric properties between healthy and unhealthy tissues in the microwave range are used for the identification of diseases like breast cancer. UWB imaging exploits the significant dielectric contrast between malignant tumors and normal breast tissue in the UWB range, which results in strong scattering [18]. The dielectric contrast, which is estimated to be greater than 2:1, is much higher than the few percent contrast in radiographic density exploited by X-ray mammography. Therefore, even though active microwave imaging does not offer the high spatial resolution provided by X-rays, it has the potential to offer improved sensitivity and specificity. In addition, active microwave imaging is a non-ionizing and noninvasive modality, and does not require breast compression [93]. Various proposals of UWB radar setups for microwave breast imaging have been presented [77, 93, 115].

Detection of water accumulation

This application is crucial for people with diseases such as pulmonary edema due to a failure of the heart or urinary incontinence. In Germany around 10 million people suffer from urinary incontinence with an annual cost of more than 1 billion Euro for their treatment [50]. In urology, neurological diseases can cause the loss of ability to determine the fluid level in the bladder (eg. paraplegic patients), which makes an external determination of the urination intervals necessary and often leads to a permanent catheterization. Thus, monitoring of the water accumulation in the bladder using a UWB radar would be of great help to people suffering from diseases such as urinary incontinence. In contrast, by monitoring the urine in the bladder of patients, a long-time empty bladder could be identified. This could help elderly people that suffer from dehydration due to their inability to feel thirsty. A radar concept that estimates the amount of water in the human body (e.g. the bladder) based on impulse radio ultra-wideband (IR-UWB) is proposed in [92]. Detection is possible due to the high difference in the dielectric permittivity of water in comparison to different living tissues (such as fat or muscle). For such applications, a high resolution is needed as provided by UWB – due to the huge bandwidth.

Stroke detection

Generally speaking, a stroke is the loss of brain functionality due to a disturbance in the blood flow to the brain. It is classified into two major categories: ischemic and hemorrhagic [59]. Contrary treatment is required immediately for both types and it is essential not only to detect a stroke, but to distinguish between the two types right after it has happened [132]. Stroke detection is regarded as a challenging issue in the medical world. Conventional imaging techniques such as magnetic resonance imaging (MRI) and computed tomography (CT) scanning are widely used to detect and classify strokes. However, besides their high cost, these techniques suffer from one major problem: they cannot be made available in the ambulance and therefore cannot be used shortly after the stroke occurs. As an alternative technique, microwave medical imaging using UWB signals has recently been considered by some research groups around the world [59, 71]. Due to its low system complexity, low cost and non-ionizing signal characteristics, UWB imaging has a high potential to complement conventional imaging systems in particular situations (e.g. emergency scenarios, medical practices).

References

[1] M. M. Abdul-Razzak, B. A. Hardwick, G. L. Hey-Shipton, P. A. Matthews, J. R. T. Monson, and R. C. Kester, "Microwave thermography for medical applications," *Physical Science, Measurement and Instrumentation, Management and Education – Reviews, IEE Proceedings A*, vol. 134, no. 2, pp. 171–74, February 1987.

[2] acam-messelectronic GmbH, "TDC-GPX: Ultra-high Performance 8 Channel Time-to-Digital Converter," http://www.acam.de/fileadmin/Download/pdf/English/DB_GPX_e.pdf, 2007.

[3] G. Adamiuk, *Methoden zur Realisierung von dual-orthogonal, linear polarisierten Antennen für die UWB-Technik*, ser. Karlsruher Forschungsberichte aus dem Institut für Hochfrequenztechnik und Elektronik; 61. Karlsruhe: KIT Scientific Publishing, 2010. [Online]. Available: http://digbib.ubka.uni-karlsruhe.de/volltexte/1000019874.

[4] G. Adamiuk, S. Beer, W. Wiesbeck, and T. Zwick, "Dual-orthogonal polarized antenna for UWB-IR technology," *IEEE Antennas and Wireless Propagation Letters*, vol. 8, pp. 981–84, 2009.

[5] G. Adamiuk, C. Heine, W. Wiesbeck, and T. Zwick, "Antenna array system for UWB-monopulse-radar," in *International Workshop on Antenna Technology, iWAT*, March 2010.

[6] G. Adamiuk, M. Janson, T. Zwick, and W. Wiesbeck, "Dual-polarized UWB antenna array," in *International Conference on Ultra-Wideband, ICUWB*, September 2009.

[7] G. Adamiuk, M. Pauli, and T. Zwick, "Principle for the realization of dual-orthogonal linearly polarized antennas for UWB techniques," *International Journal of Antennas and Propagation*, 2011.

[8] G. Adamiuk, J. Timmermann, C. Roblin, W. Dullaert, P. Gentner, K. Witrisal, T. Fügen, O. Hirsch, and G. Shen, "Chapter 6: RF Aspects in Ultra-WideBand Technology," in *Verdone, R. and Zanella, A.: Pervasive Mobile and Ambient Wireless Communications, COST Action 2100*. Springer, 2012, pp. 249–300.

[9] G. Adamiuk, J. Timmermann, W. Wiesbeck, and T. Zwick, "A novel concept of a dual-orthogonal polarized ultra wideband antenna for medical applications," in *3rd European Conference on Antennas and Propagation, EuCAP*, March 2009.

[10] G. Adamiuk, W. Wiesbeck, and T. Zwick, "Multi-mode antenna feed for ultra wideband technology," in *IEEE Radio and Wireless Symposium, RWS*, January 2009.

[11] G. Adamiuk, T. Zwick, and W. Wiesbeck, "Dual-orthogonal polarized Vivaldi antenna for ultra wideband applications," in *17th International Conference on Microwaves, Radar and Wireless Communications, MIKON*, May 2008.

[12] G. Adamiuk, T. Zwick, and W. Wiesbeck, "Compact, Dual-Polarized UWB-Antenna, Embedded in a Dielectric," *IEEE Transactions on Antennas and Propagation*, vol. 58, pp. 279–86, February 2010.

[13] G. Adamiuk, T. Zwick, and W. Wiesbeck, "UWB antennas for communication systems," *Proceedings of the IEEE*, vol. 100, pp. 2308–21, 2012.

[14] G. Adamiuk, L. Żwirełło, S. Beer, and T. Zwick, "Omnidirectional dual-orthogonal polarized UWB antenna," in *European Microwave Week, EuMW*, September 2010.

[15] Agilent Technology, "Advanced Design System (ADS)," http://www.home.agilent.com/en/pc-1297113/advanced-design-system-ads, 2009.

[16] B. Ahmed and M. Ramon, "Coexistence between UWB and other communication systems – tutorial review," *International Journal Ultra Wideband Communications and Systems*, vol. 1, no. 1, pp. 67–80, 2009.

[17] O. Albert and C. Mecklenbräuker, "An 8-bit programmable fine delay circuit with step size 65 ps for an ultrawideband pulse position modulation testbed," in *15th European Signal Processing Conference*, September 2007.

[18] B. Allen and M. Dohler, *Ultra-wideband antennas and propagation for communications, radar and imaging*. Arizona State University: Wiley, 2007.

[19] M. Anis and R. Tielert, "Design of UWB pulse radio transceiver using statistical correlation technique in frequency domain," in *Advances in Radio Science – An Open Access Journal of the U.R.S.I. Landesausschuss in der Bundesrepublik Deutschland e.V.*, pp. 297–304, 2007.

[20] C. Ascher, L. Żwirełło, T. Zwick, and G. Trommer, "Integrity monitoring for UWB/INS tightly coupled pedestrian indoor scenarios," in *International Conference on Indoor Positioning and Indoor Navigation, IPIN*, September 2011.

[21] S. Bagga, L. Zhang, W. Serdijn, J. Long, and E. Busking, "A quantized analog delay for an IR-UWB quadrature downconversion autocorrelation receiver," in *IEEE International Conference on Ultra-Wideband, ICU*, September 2005.

[22] P. Bahl and V. N. Padmanabhan, "RADAR: an in-building RF-based user location and tracking system," *IEEE 19th Annual Joint Conference of the IEEE Computer and Communications Societies*, vol. 2, pp. 775–84, 2000.

[23] C. Balanis, *Advanced Engineering Electromagnetics*. New York: Wiley, 1989.

[24] C. Balanis, *Antenna Theory: Analysis and Design*. Wiley-Interscience, 2005.

[25] N. S. Barker and G. M. Rebeiz, "Distributed MEMS true-time delay phase shifters and wide-band switches," *IEEE Transactions on Microwave Theory and Techniques*, vol. 46, issue 11, part 2, pp. 1881–90, 1998.

[26] N. Behdad and K. Sarabandi, "A compact antenna for ultrawide-band applications," *IEEE Transactions on Antennas and Propagation*, vol. 53, pp. 2185–92, July 2005.

[27] H. Booker, "Slot aerials and their relation to complementary wire aerials (Babinet's principle)," *Journal of the Institution of Electrical Engineers – Part IIIA: Radiolocation*, vol. 93, pp. 620–26, 1946.

[28] B. H. Burdine, "The spiral antenna," Massachusetts Institute of Technology, Research Lab. Tech. Rep., April 1955.

[29] M. Cavallaro, E. Ragonese, and G. Palmisano, "An ultra-wideband transmitter based on a new pulse generator," in *IEEE Radio Frequency Integrated Circuits Symposium, RFIC*, April 2008.

[30] S. Chang, "CMOS 5th derivative Gaussian impulse generator for UWB application," Master's thesis, Graduate School of University of Texas, Arlington, December 2005.

[31] S. Chang, S. Jung, S. Tjuatja, J. Gao, and Y. Joo, "A CMOS 5th derivative impulse generator for an IR-UWB," in *49th International Midwest Symposium on Circuits and Systems (MWSCAS)*, August 2006.

[32] Y. Chao and R. Scholtz, "Optimal and suboptimal receivers for ultra-wideband transmitted reference systems," in *IEEE Global Telecommunications Conference, GLOBECOM*, December 2003.

[33] Z. Chen and Y. Zhang, "A modified synchronization scheme for impulse-based UWB," in *6th International Conference on Information, Communications and Signal Processing*, December 2007.

[34] C. C. Chong, F. Watanabe, and H. Inamura, "Potential of UWB technology for the next generation wireless communications," in *IEEE Ninth International Symposium on Spread Spectrum Techniques and Applications*, pp. 422–29, August 2006.

[35] A. Christ, A. Klingenböck, and N. Kuster, "Exposition durch koerpernahe Sender im Rumpfbereich, Arbeitspaket I: Bestandsaufnahme," Foundation for Research on Information Technologies in Society, Swiss Federal of Technology, ETHZ, Zurich, Tech. Rep., 2004.

[36] D. C. Daly, P. P. Mercier, M. Bhardwaj, A. L. Stone, Z. N. Aldworth, T. L. Daniel, J. Voldman, J. G. Hildebrand, and A. P. Chandrakasan, "A pulsed UWB receiver SoC for insect motion control," *IEEE Journal of Solid-State Circuits*, vol. 45, pp. 153–66, 2010.

[37] P. K. Datta, X. Fan, and G. Fischer, "A transceiver front-end for ultra-wide-band applications," *IEEE Transactions on Circuits and Systems II: Express Briefs*, vol. 54, pp. 362–66, 2007.

[38] A. De Angelis, M. Dionigi, A. Moschitta, R. Giglietti, and P. Carbone, "Characterization and modeling of an experimental UWB pulse-based distance measurement system," *IEEE Transactions on Instrumentation and Measurement*, vol. 58, pp. 1479–86, May 2009.

[39] G. Deschamps, "Impedance properties of complementary multiterminal planar structures," *IRE Transactions on Antennas and Propagation*, vol. 7, pp. 371–78, December 1959.

[40] S. Duenas, "Design of a DS-UWB transmitter," Master's thesis, KTH Stockholm, March 2005.

[41] H. Dunger, "World-wide regulation and standardisation overview," in *Integrated Project EUWB*, http://www.euwb.eu/deliverables/EUWB_D9.1_v1.0_2008-09-15.pdf, September 2009.

[42] J. Dyson, "The equiangular spiral antenna," *IRE Transactions on Antennas and Propagation*, vol. 7, pp. 181–87, April 1959.

[43] M. Eisenacher, "Optimierung von Ultra-Wideband-Signalen (UWB)," PhD dissertation, Forschungsberichte aus dem Institut für Nachrichtentechnik der Universität Karlsruhe (TH), August 2006.

[44] European Commission, "Commission decision on allowing the use of the radio spectrum for equipment using ultra-wideband technology in a harmonised manner in the community," *Official Journal of the European Union*, vol. 55, February 2007.

[45] E. G. Farr and C. E. Baum, "Time domain characterization of antennas with TEM feeds," *Sensor and Simulation Notes*, vol. 426, pp. 1–16, October 1998.

[46] Federal Communications Commission and others, "Revision of part 15 of the commission's rules regarding ultra-wideband transmission systems," *ET Docket 98–153. FCC 02-48*, 2002.

[47] J. Ferlay, P. Autier, M. Boniol, M. Heanue, M. Colombet, and P. Boyle, "Estimates of the cancer incidence and mortality in Europe in 2006," *Ann Oncol.*, vol. 18, no. 3, pp. 581–92, 2007.

[48] G. Fischer, O. Klymenko, D. Martynenko, and H. Luediger, "An impulse radio UWB transceiver with high-precision TOA measurement unit," in *International Conference on Indoor Positioning and Indoor Navigation, IPIN*, September 2010.

[49] A. Fort, C. Desset, P. De Doncker, P. Wambacq, and L. Van Biesen, "An ultra-wideband body area propagation channel model – from statistics to implementation," *IEEE Transactions on Microwave Theory and Techniques*, vol. 54, pp. 1820–26, June 2006.

[50] R. E. Fromm, J. Varon, and L. Gibbs, "Congestive heart failure and pulmonary edema for the emergency physician," *Journal of Emergency Medicine*, vol. 13, pp. 71–87, 1995.

[51] T. Fügen, J. Maurer, T. Kayser, and W. Wiesbeck, "Capability of 3-D ray tracing for defining parameter sets for the specification of future mobile communications systems," *IEEE Transactions on Antennas and Propagation*, vol. 54, pp. 3125–37, November 2006.

[52] C. Gabriel, S. Gabriel, and E. Corthout, "The dielectric properties of biological tissues: I. Literature survey," *Phys. Med. Biol.*, vol. 41, pp. 2231–49, November 1996.

[53] S. Gabriel, R. W. Lau, and C. Gabriel, "The dielectric properties of biological tissues: II. Measurements in the frequency range 10 Hz to 20 GHz," *Phys. Med. Biol.*, vol. 41, pp. 2251–69, November 1996.

[54] S. Gabriel, R. W. Lau, and C. Gabriel, "The dielectric properties of biological tissues: III. Parametric models for the dielectric spectrum of tissues," *Phys. Med. Biol.*, vol. 41, pp. 2271–93, November 1996.

[55] N. Geng and W. Wiesbeck, *Planungsmethoden für die Mobilkommunikation – Funknetzplanung unter realen physikalischen Ausbreitungsbedingungen*. Springer, 1998.

[56] S. Gezici, "A survey on wireless position estimation," *Wireless Personal Communications*, vol. 44, pp. 263–82, 2007.

[57] S. Gezici, Z. Tian, G. Giannakis, H. Kobayashi, A. Molisch, H. Poor, and Z. Sahinoglu, "Localization via ultra-wideband radios: a look at positioning aspects for future sensor networks," *IEEE Signal Processing Magazine*, vol. 22, pp. 70–84, July 2005.

[58] E. Gschwendtner and W. Wiesbeck, "Ultra-broadband car antennas for communications and navigation applications," *IEEE Transactions on Antennas and Propagation*, vol. 51, pp. 2020–27, August 2003.

[59] M. Guardiola, L. Jofre, and J. Romeu, "3D UWB tomography for medical imaging applications," in *IEEE Antennas and Propagation Society International Symposium, APSURSI*, July 2010.

[60] M. Hamalainen, A. Taparugssanagorn, R. Tesi, and J. Iinatti, "Wireless medical communications using UWB," in *IEEE International Conference on Ultra-Wideband (ICUWB)*, September 2009.

[61] S. Hantscher, "Comparison of UWB target identification algorithms for through-wall imaging applications," in *3rd European Radar Conference, EuRAD*, September 2006.

[62] S. Hantscher, A. Reisenzahn, and C. Diskus, "Analysis of imaging radar algorithms for the identification of targets by their surface shape," in *IEEE International Conference on Ultra-Wideband, ICUWB*, October 2006.

[63] C. Harrison, and C. Williams, "Transients in wide-angle conical antennas," *IEEE Transactions on Antennas and Propagation*, vol. 13, pp. 236–46, March 1965.

[64] J. Hightower and G. Borriello, "Location systems for ubiquitous computing," *Computer*, vol. 34, pp. 57–66, August 2001.

[65] Hittite, "Wideband LNA module HMC-C022," http://www.hittite.com/content/documents/data_sheet/hmc-c022.pdf, 2012.

[66] R. Hoctor and H. Tomlinson, "Delay-hopped transmitted-reference RF communications," in *IEEE Conference on Ultra Wideband Systems and Technologies*, 2002.

[67] *IEEE Std 149-1979: IEEE Standard Test Procedures for Antennas*, IEEE, Institute of Electrical and Electronics Engineers, 1979.

[68] *IEEE Std 145-1993:IEEE Standard Definitions of Terms for Antennas*, IEEE, Institute of Electrical and Electronics Engineers, 1993.

[69] IHP, "http://www.ihp-microelectronics.com/en/services/mpw-prototyping/sigec-bicmos-technologies.html," 2013.

[70] I. Immoreev and T. H. Tao, "UWB radar for patient monitoring," *IEEE Aerospace and Electronic Systems Magazine*, vol. 23, pp. 11–18, November 2008.

[71] M. Jalilvand, T. Zwick, W. Wiesbeck, and E. Pancera, "UWB synthetic aperture-based radar system for hemorrhagic head-stroke detection," in *Radar Conference (RADAR)*, May 2011.

[72] M. Janson, J. Pontes, T. Fuegen, and T. Zwick, "A hybrid deterministic-stochastic propagation model for short-range MIMO-UWB communication systems," *FREQUENZ*, vol. 66, no. 7-8, pp. 193–203, 2012.

[73] A. Jha, R. Gharpurey, and P. Kinget, "A 3 to 5-GHz UWB pulse radio transmitter in 90 nm CMOS," in *IEEE Radio Frequency Integrated Circuits Symposium, RFIC*, April 2008.

[74] E. B. Joy and D. T. Paris, "A practical method for measuring the complex polarization ratio of arbitrary antennas," *IEEE Transactions on Antennas and Propagation*, vol. 21, pp. 432–35, March 1973.

[75] T. Kaiser, F. Zheng, and E. Dimitrov, "An overview of ultra-wide-band systems with MIMO," *Proceedings of the IEEE*, vol. 97, pp. 285–312, February 2009.

[76] P. Keranen, K. Maatta, and J. Kostamovaara, "Wide-range time-to-digital converter with 1-ps single-shot precision," *IEEE Transactions on Instrumentation and Measurement*, vol. 60, no. 9, pp. 3162–72, September 2011.

[77] M. Klemm, I. Craddock, J. Leendertz, A. Preece, and R. Benjamin, "Radar-based breast cancer detection using a hemispherical antenna array – experimental results," *IEEE Transactions on Antennas and Propagation*, vol. 57, pp. 1692–704, June 2009.

[78] O. Klymenko, G. Fischer, and D. Martynenko, "A high band non-coherent impulse radio UWB receiver," in *IEEE International Conference on Ultra-Wideband, ICUWB*, 2008.

[79] J. Kolakowski, "Application of ultra-fast comparator for UWB pulse time of arrival measurement," in *IEEE International Conference on Ultra-Wideband, ICUWB*, September 2011.

[80] T. Kürner, M. Jacob, R. Piesiewicz, and J. Schöbel, "An integrated simulation environment for the investigation of future THz communication systems," in *International Symposium on Performance Evaluation of Computer and Telecommunication Systems (SPECTS)*, July 2007.

[81] A. Kuthi, M. Behrend, T. Vernier, and M. Gundersen, "Bipolar nanosecond pulse generation using transmission lines for cell electro-manipulation," in *26th International Power Modulator Symposium*, May 2004.

[82] D. H. Kwon, "Effect of antenna gain and group delay variations on pulse-preserving capabilities of ultrawideband antennas," *IEEE Transactions on Antennas and Propagation*, vol. 54, pp. 2208–15, August 2006.

[83] D. Lachartre, B. Denis, D. Morche, L. Ouvry, M. Pezzin, B. Piaget, J. Prouvee, and P. Vincent, "A 1.1nJ/b 802.15.4a-compliant fully integrated UWB transceiver in 0.13 μm CMOS," in *IEEE International Solid-State Circuits Conference – Digest of Technical Papers, ISSCC*, 2009.

[84] R. Lakes, H. S. Yoon, and J. L. Katz, "Ultrasonic wave propagation and attenuation in wet bone," *Journal of Biomedical Engineering*, vol. 8, pp. 143–48, April 1986. [Online]. Available: http://www.sciencedirect.com/science/article/pii/014154258690049X.

[85] A. Lambrecht, P. Laskowski, S. Beer, and T. Zwick, "Frequency invariant beam steering for short-pulse systems with a Rotman lens," *International Journal of Antennas and Propagation*, 2010.

[86] J. D. D. Langley, P. S. Hall, and P. Newham, "Novel ultrawide-bandwidth Vivaldi antenna with low crosspolarisation," *Electronic Letters*, vol. 29, no. 23, 1993.

[87] A. Lecointre, D. Dragomirescu, and R. Plana, "Channel capacity limitations versus hardware implementation for UWB impulse radio communications," *CoRR*, vol. abs/1002.0574, 2010. [Online]. Available: http://dblp.uni-trier.de/db/journals/corr/corr1002.html#abs-1002-0574.

[88] W. Lee, S. Kunaruttanapruk, and S. Jitapunkul, "Optimal pulse shape design for UWB systems with timing jitter," *IEICE Transactions on Communications*, vol. E91-B, no. 3, pp. 772–83, March 2008.

[89] M. Leib, W. Menzel, B. Schleicher, and H. Schumacher, "Vital signs monitoring with a UWB radar based on a correlation receiver," in *IEEE European Conference on Antennas and Propagation, EuCAP*, April 2010.

[90] X. Li, "Anwendung von dual-orthogonal polarisierten Antennen in UWB-Imaging-Systemen," Master's thesis, Karlsruhe Institute of Technology, May 2009.

[91] X. Li, G. Adamiuk, M. Janson, and T. Zwick, "Polarization diversity in ultra-wideband imaging systems," in *International Conference on Ultra Wideband, ICUWB*, September 2010.

[92] X. Li, G. Adamiuk, E. Pancera, and T. Zwick, "Physics-based propagation characterisations of UWB signals for the urine detection in human bladder," *International Journal on Ultra Wideband Communications and Systems*, vol. 2, pp. 94–103, December 2011.

[93] X. Li, S. K. Davis, S. C. Hagness, D. W. van der Weide, and B. D. Van Veen, "Microwave imaging via space-time beamforming: experimental investigation of tumor detection in multilayer breast phantoms," *IEEE Transactions on Microwave Theory and Techniques*, vol. 52, pp. 1856–65, August 2004.

[94] X. Li, L. Żwirełło, M. Jalilvand, and T. Zwick, "Design and near-field characterization of a planar on-body UWB slot-antenna for stroke detection," in *IEEE International Workshop on Antenna Technology, iWAT*, March 2012.

[95] G. Lim, Y. Zheng, W. Yeoh, and Y. Lian, "A novel low power UWB transmitter IC," in *IEEE Radio Frequency Integrated Circuits (RFIC) Symposium*, June 2006.

[96] S. Lin and T. Chiueh, "Performance analysis of impulse radio under timing jitter using M-ary bipolar pulse waveform and position modulation," in *IEEE Conference on Ultra Wideband Systems and Technologies*, November 2003.

[97] H. Liu, H. Darabi, P. Banerjee, and J. Liu, "Survey of wireless indoor positioning techniques and systems," *IEEE Transactions on Systems, Man, and Cybernetics, Part C: Applications and Reviews*, vol. 37, pp. 1067–80, November 2007.

[98] D. Lochmann, *Digitale Nachrichtentechnik*. Verlag Technik Berlin, 1995.

[99] D. Martynenko, G. Fischer, and O. Klymenko, "A high band impulse radio UWB transmitter for communication and localization," in *IEEE International Conference on Ultra-Wideband, ICUWB*, 2009.

[100] "Maxima, a Computer Algebra System," http://maxima.sourceforge.net, 2012.

[101] P. Mayes, "Frequency-independent antennas and broad-band derivatives thereof," *Proceedings of the IEEE*, vol. 80, pp. 103–12, January 1992.

[102] C. Mensing and S. Plass, "Positioning algorithms for cellular networks using TDOA," in *IEEE International Conference on Acoustics, Speech and Signal Processing, ICASSP*, May 2006.

[103] P. P. Mercier, D. C. Daly, and A. P. Chandrakasan, "A 19pJ/pulse UWB transmitter with dual capacitively-coupled digital power amplifiers," in *IEEE Radio Frequency Integrated Circuits Symposium, RFIC*, April 2008.

[104] S. M. Metev and V. P. Veiko, *Laser Assisted Microtechnology*, 2nd ed. Berlin, Germany: Springer, 1998.

[105] R. Meys, "A summary of the transmitting and receiving properties of antennas," *IEEE Antennas and Propagation Magazine*, vol. 42, pp. 49–53, June 2000.

[106] E. K. Miller and F. J. Deadrick, "Visualizing near-field energy flow and radiation," *IEEE Antennas and Propagation Magazine*, vol. 42, pp. 46–54, December 2000.

[107] W. Mitchell, "Avalanche transistors give fast pulses," in *Electronic Design*, 1968.

[108] A. F. Molisch, *Wireless communications*, 2nd ed. Chichester: Wiley, 2011.

[109] A. Molisch, K. Balakrishnan, D. Cassioli, C. Chong, S. Emami, A. Fort, J. Karedal, J. Kunisch, H. Schantz, U. Schuster, and K. Siwiak, "IEEE 802.15.4a channel model – final report," IEEE 802.15-04-0662-00-0004a, San Antonio, Texas, USA, Tech. Rep., November 2004.

[110] A. F. Molisch, J. R. Foerster, and M. Pendergrass, "Channel models for ultrawideband personal area networks," *IEEE Wireless Communications*, vol. 10, pp. 14–21, December 2003.

[111] C. Müller, S. Zeisberg, H. Seidel, and A. Finger, "Spreading properties of time hopping codes in ultra wideband systems," in *IEEE 7th Symposium on Spread-Spectrum Techniques and Applications*, September 2002.

[112] S. A. Z. Murad, R. K. Pokharel, A. I. A. Galal, R. Sapawi, H. Kanaya, and K. Yoshida, "An excellent gain flatness 3.0–7.0 GHz CMOS PA for UWB applications," *Microwave and Wireless Components Letters, IEEE*, vol. 20, no. 9, pp. 510–12, 2010.

[113] Y. Mushiake, "Self-complementary antennas," *IEEE Antennas and Propagation Magazine*, vol. 34, pp. 23–29, December 1992.

[114] M. Neinhus, S. Held, and K. Solobach, "FIR-filter based equalization of ultra wideband mutual coupling on linear antenna arrays," in *2nd International ITG Conference on Antennas, INICA*, 2007.

[115] N. K. Nikolova, "Microwave imaging for breast cancer," *IEEE Microwave Magazine*, vol. 12, pp. 78–94, December 2011.

[116] R. Nilavalan, I. J. Craddock, A. Preece, J. Leendertz, and R. Benjamin, "Wideband Microstrip Patch Antenna Design for Breast Cancer Tumour Detection," *IET Antennas Propagation Microwaves*, vol. 1, no. 2, pp. 277–81, April 2007.

[117] T. Norimatsu, R. Fujiwara, M. Kokubo, M. Miyazaki, A. Maeki, Y. Ogata, S. Kobayashi, N. Koshizuka, and K. Sakamura, "A UWB-IR transmitter with digitally controlled pulse generator," *IEEE Journal of Solid-State Circuits*, vol. 42, pp. 1300–09, June 2007.

[118] U. Onunkwo, "Timing jitter in ultra wideband (UWB) systems," PhD dissertation, School of Electrical and Computer Engineering, Georgia Institute of Technology, May 2006.

[119] A. Oppenheim, *Discrete-Time Signal Processing*. Prentice Hall, Inc., 1989.

[120] I. Oppermann, M. Hämäläinen, and J. Iinatti, *UWB Theory and Applications*. J. Wiley & Sons, 2006.

[121] J. Padgett, J. Koshy, and A. Triolo, "Physical-layer modeling of UWB interference effects," Wireless Systems and Networks Research, Telcordia Technologies Inc., Arlington, Tech. Rep., 2003.

[122] K. Pahlavan, X. Li, and J. Mäkelä, "Indoor geolocation science and technology," *IEEE Communications Magazine*, vol. 40, no. 2, pp. 112–18, February 2002.

[123] E. Pancera, *Strategies for time domain characterization of UWB components and systems*, ser. Karlsruher Forschungsberichte aus dem Institut für Hochfrequenztechnik und Elektronik; 57. Karlsruhe: Universitätsverlag, 2009. [Online]. Available: http://digbib.ubka. uni-karlsruhe.de/volltexte/1000012414.

[124] E. Pancera and W. Wiesbeck, "Correlation properties of the pulse transmitted by UWB antennas," in *International Conference on Electromagnetics in Advanced Applications, ICEAA*, September 2009.

[125] E. Pancera, T. Zwick, and W. Wiesbeck, "Correlation properties of UWB radar target impulse responses," in *IEEE Radar Conference, RadarCon*, May 2009.

[126] E. Pancera, T. Zwick, and W. Wiesbeck, "Full polarimetric time domain calibration for UWB radar systems," in *European Radar Conference, EuRAD 2009*, October 2009.

[127] E. Pancera, T. Zwick, and W. Wiesbeck, "Spherical fidelity patterns of UWB antennas," *IEEE Transactions on Antennas and Propagation*, vol. 59, pp. 2111–19, June 2011.

[128] R. Pantoja, A. Sapienza, and F. Filho, "A microwave printed planar log-periodic dipole array antenna," *IEEE Transactions on Antennas and Propagation*, vol. 35, pp. 1176–78, October 1987.

[129] S. Paquelet, L. Aubert, and B. Uguen, "An impulse radio asynchronous transceiver for high data rates," in *Conference on Ultrawideband Systems and Technologies*, September 2004.

[130] C. N. Paulson, J. T. Chang, C. E. Romero, J. Watson, F. J. Pearce, and N. Levin, "Ultra-wideband radar methods and techniques of medical sensing and imaging," in *SPIE International Symposium on Optics*, October 2005.

[131] C. Peixeiro, "Design of log-periodic dipole antennas," *IEE Proceedings Microwaves, Antennas and Propagation*, vol. 135, pp. 98–102, April 1988.

[132] M. Persson, "UWB in medical diagnostics and treatment," in *IEEE-APS Topical Conference on Antennas and Propagation in Wireless Communications, APWC*, September 2011.

[133] A. Phan, J. Lee, V. Krizhanovskii, Q. Le, S.-K. Han, and S.-G. Lee, "Energy-efficient low-complexity CMOS pulse generator for multiband UWB impulse radio," *IEEE Transactions on Circuits and Systems I: Regular Papers*, vol. 55, pp. 3552–63, December 2008.

[134] M. Porebska, G. Adamiuk, C. Sturm, and W. Wiesbeck, "Accuracy of algorithms for UWB localization in NLOS scenarios containing arbitrary walls," in *The Second European Conference on Antennas and Propagation, EuCAP*, November 2007.

[135] M. Porebska, T. Kayser, and W. Wiesbeck, "Verification of a hybrid ray-tracing/FDTD model for indoor ultra-wideband channels," in *European Conference on Wireless Technologies*, October 2007.

[136] D. Pozar, *Microwave Engineering*. John Wiley, second edition, ISBN 0-471-17096-8, 1998.

[137] P. Prasithsangaree, P. Krishnamurthy, and P. Chrysanthis, "On indoor position location with wireless LANs," in *13th IEEE International Symposium on Personal, Indoor and Mobile Radio Communications*, September 2002.

[138] S. Promwong and J. Takada, "Free space link budget estimation scheme for ultra wideband impulse radio with imperfect antennas," *IEICE Electronics Express*, vol. 1, pp. 188–92, 2004.

[139] A. Rabbachin, "Low complexity UWB receivers with ranging capabilities," PhD dissertation, Faculty of Technology, Department of Electrical and Information Engineering, Centre for Wireless Communications, University of Oulu, Finland, March 2008.

[140] A. Rabbachin, J. Montillet, P. Cheong, G. De Abreu, and I. Oppermann, "Non-coherent energy collection approach for TOA estimation in UWB systems," in *14th IST Mobile and Wireless Communications Summit*, June 2005.

[141] J. Reed, *An Introduction to Ultra Wideband Communication Systems*. Prentice Hall Communications Engineering and Emerging Technologies Series, 2005.

[142] J. Reed, *An Introduction to Ultra Wideband Communication Systems*, 1st ed. Upper Saddle River, NJ, USA: Prentice Hall Press, 2005.

[143] A. Reisenzahn, "Hardwarekomponenten für Ultra-Wideband Radio," Master's thesis, Institut für Nachrichtentechnik/Informationstechnik, University of Linz, Austria, 2003.

[144] H. Rohling, Ed., *OFDM: Concepts for Future Communication Systems*. Wiesbaden: Springer, 2011.

[145] Z. Rudolf, S. Juergen, and T. Reiner, "Imaging of propagation environment by channel sounding," in *XXVIIIth General Assembly of URSI*, October 2005.

[146] V. Rumsey, *Frequency independent antennas*. Electrical science series. Academic Press, 1966.

[147] Z. Sahinoglu, S. Gezici, and I. Güvenc, *Ultra-wideband Positioning Systems: Theoretical Limits, Ranging Algorithms, and Protocols*. Cambridge University Press, 2008.

[148] S. Sato and T. T. Kobayashu, "Path-loss exponents of ultra wideband signals in line-of-sight environments," in *In Proceedings of the IEEE 8th International Symposium on Spread Spectrum Techniques and Applications*, pp. 488–92, September 2004.

[149] H. G. Schantz, "A brief history of UWB antennas," *IEEE Aerospace and Electronic Systems Magazine*, vol. 19, pp. 22–26, April 2004.

[150] D. Schaubert, E. Kollberg, T. Korzeniowski, T. Thungren, J. Johansson, and K. Yngvesson, "Endfire tapered slot antennas on dielectric substrates," *IEEE Transactions on Antennas and Propagation*, vol. 33, pp. 1392–1400, December 1985.

[151] B. Scheers, M. Acheroy, and A. V. Vorst, "Time-domain simulation and characterisation of TEM horns using a normalised impulse response," *IEE Proceedings – Microwaves, Antennas Propagation*, vol. 147, pp. 463–68, December 2000.

[152] B. Schleicher, J. Dederer, M. Leib, I. Nasr, A. Trasser, W. Menzel, and H. Schumacher, "Highly compact impulse UWB transmitter for high-resolution movement detection," in *IEEE International Conference on Ultra-Wideband, ICUWB*, September 2008.

[153] I. Sharp, K. Yu, and Y. J. Guo, "GDOP analysis for positioning system design," *IEEE Transactions on Vehicular Technology*, vol. 58, pp. 3371–82, September 2009.

[154] A. Shlivinski, E. Heyman, and R. Kastner, "Antenna characterization in the time domain," *IEEE Transactions on Antennas and Propagation*, vol. 45, pp. 1140–49, July 1997.

[155] B. Sklar, *Digital Communications – Fundamentals and Applications*, 2nd ed. Prentice Hall, ISBN 0-13-084788-7, 2000.

[156] M. I. Skolnik, *Introduction to Radar Systems*. New York: McGraw-Hill, 1980.

[157] A. A. Smith, "Received voltage versus antenna height," *IEEE Transactions on Electromagnetic Compatibility*, vol. EMC-11, pp. 104–11, August 1969.

[158] W. Sörgel, *Charakterisierung von Antennen für die Ultra-Wideband-Technik*, ser. Forschungsberichte aus dem Institut für Höchstfrequenztechnik und Elektronik der Universität Karlsruhe (TH); 51. IHE, 2007. [Online]. Available: http://digbib.ubka.uni-karlsruhe.de/volltexte/1000007210.

[159] W. Sörgel and W. Wiesbeck, "Influence of the antennas on the ultra-wideband transmission," *EURASIP Journal on Advances in Signal Processing*, pp. 296–305, 2005.

[160] E. Staderini, "UWB radars in medicine," *IEEE Aerospace and Electronic Systems Magazine*, vol. 17, pp. 13–18, January 2002.

[161] L. Stoica, "Non-coherent energy detection transceivers for ultra wideband impulse radio systems," PhD dissertation, Faculty of Technology, Department of Electrical and Information Engineering, University of Oulu, 2008, ISBN 978-951-42-8717-6.

[162] L. Stoica and I. Oppermann, "Modelling and simulation of a non-coherent IR UWB transceiver architecture with TOA estimation," in *17th IEEE International Symposium on Personal, Indoor and Mobile Radio Communications (PIMRC)*, September 2006.

[163] M. L. Stowell, B. J. Fasenfest, and D. A. White, "Investigation of radar propagation in buildings: A 10-billion element cartesian-mesh FDTD simulation," *IEEE Transactions on Antennas and Propagation*, vol. 56, no. 8, pp. 2241–50, 2008.

[164] A. Tamtrakarn, H. Ishikuro, K. Ishida, M. Takamiya, and T. Sakurai, "A 1-V 299μW flashing UWB transceiver based on double thresholding scheme," in *Symposium on VLSI Circuits, Digest of Technical Papers*, 2006.

[165] J.-Y. Tham, B. L. Ooi, and M. Leong, "Diamond-shaped broadband slot antenna," in *IEEE International Workshop on Antenna Technology: Small Antennas and Novel Metamaterials, IWAT*, March 2005.

[166] J. Y. Tham, B. L. Ooi, and M. S. Leong, "Novel design of broadband volcano-smoke antenna," in *IEEE Antennas and Propagation Society International Symposium*, July 2005.

[167] R. S. Thomä, H.-I. Willms, T. Zwick, R. Knöchel, and J. Sachs, Eds., *UKoLoS Ultra-Wideband Radio Technologies for Communications, Localization and Sensor Applications*. Intech, September 2012.

[168] J. Timmermann, *Systemanalyse und Optimierung der Ultrabreitband-Übertragung*, ser. Karlsruher Forschungsberichte aus dem Institut für Hochfrequenztechnik und Elektronik; 58. Karlsruhe: KIT Scientific Publishing, 2010. [Online]. Available: http://digbib.ubka.uni-karlsruhe.de/volltexte/1000014984.

[169] J. Timmermann, P. Walk, A. Rashidi, W. Wiesbeck, and T. Zwick, "Compensation of a non-ideal UWB antenna performance," *Frequenz, Journal of RF-Engineering and Telecommunications*, vol. 63, pp. 183–86, 2009.

[170] Ubisense Group, "Ubisense series 7000 IP rated sensor," http://www.ubisense.net/en/media/pdfs/factsheets_pdf/56505_ubisense-series-7000-ip-rated-sensor-en090624.pdf, 2009.

[171] N. Van Helleputte and G. Gielen, "A 70 pJ/pulse analog front-end in 130 nm CMOS for UWB impulse radio receivers," *IEEE Journal of Solid-State Circuits*, vol. 44, pp. 1862–71, 2009.

[172] M. Verhelst and W. Dehaene, "Analysis of the QAC IR-UWB receiver for low energy, low data-rate communication," *IEEE Transactions on Circuits and Systems I: Regular Papers*, vol. 55, pp. 2423–32, September 2008.

[173] H. J. Visser, *Array and Phased Array Antenna Basics*. John Wiley & Sons, 2005.

[174] X. Wang, A. Young, K. Philips, and H. de Groot, "Clock accuracy analysis for a coherent IR-UWB system," in *IEEE International Conference on Ultra-Wideband (ICUWB)*, 2011.

[175] D. Ward, "No more breast cancer campaign," http://www.nomorebreastcancer.org.uk/index.html, 2008.

[176] X. Wei, K. Saito, M. Takahashi, and K. Ito, "Performances of an Implanted Cavity Slot Antenna Embedded in the Human Arm," *IEEE Transactions on Antennas and Propagation*, vol. 57, no. 4, pp. 894–99, April 2009.

[177] Wentzloff, "Pulse-based ultra-wideband transmitters for digital communication," Department of Electrical Engineering and Computer Science, Massachusetts Institute of Technology (MIT), Tech. Rep., June 2007.

[178] D. Werner, R. Haupt, and P. Werner, "Fractal antenna engineering: the theory and design of fractal antenna arrays," *IEEE Antennas and Propagation Magazine*, vol. 41, pp. 37–58, October 1999.

[179] W. Wiesbeck, G. Adamiuk, and C. Sturm, "Basic properties and design principles of UWB antennas," *Proceedings of the IEEE*, vol. 97, pp. 372–85, February 2009.

[180] W. Wiesbeck and F. Jondral, *Ultra-Wide-Band Kommunikationssysteme – Skriptum zum CCG Seminar DK 2.15*. University of Karlsruhe, 2006.

[181] M. Win and R. Scholtz, "Ultra-wide bandwidth time-hopping spread-spectrum impulse radio for wireless multiple-access communications," *IEEE Transactions on Communication*, vol. 48, pp. 679–89, April 2000.

[182] Z. Wu, F. Zhu, and C. R. Nassar, "High performance ultra-wide bandwidth systems via novel pulse shaping and frequency domain processing," in *IEEE Conference on Ultra Wideband Systems and Technologies*, pp. 53–58, 2002.

[183] Z. Xiao, G. H. Tan, R. F. Li, and K. C. Yi, "A joint localization scheme based on IR-UWB for sensor network," in *International Conference on Wireless Communications, Networking and Mobile Computing, WiCOM*, September 2011.

[184] L. Yang and G. Giannakis, "Ultra-wideband communications: An idea whose time has come, 21(6)," in *IEEE Signal Processing Magazine*, pp. 26–54, December 2004.

[185] T. Yang, S. Y. Suh, R. Nealy, W. A. Davis, and W. L. Stutzman, "Compact antennas for UWB applications," *IEEE Aerospace and Electronic Systems Magazine*, vol. 19, pp. 16–20, May 2004.

[186] R. Ye and H. Liu, "UWB TDOA localization system: Receiver configuration analysis," in *International Symposium on Signals Systems and Electronics, ISSSE*, September 2010.

[187] X. Zeng, A. Fhager, P. Persson, P. Linner, and H. Zirath, "Accuracy evaluation of ultra-wideband time domain systems for microwave imaging," *IEEE Transactions on Antennas and Propagation*, vol. 59, pp. 4279–85, November 2011.

[188] R. Zetik, J. Sachs, and R. S. Thomä, "UWB short-range radar sensing," *IEEE Instrumentation and Measurement Magazine*, vol. 10, pp. 39–45, April 2007.

[189] F. Zhang, A. Jha, R. Gharpurey, and P. Kinget, "An agile, ultra-wideband pulse radio transceiver with discrete-time wideband-IF," *IEEE Journal of Solid-State Circuits*, vol. 44, pp. 1336–51, 2009.

[190] K. Zhang and D. Li, *Electromagnetic Theory for Microwaves and Optoelectronics*, 2nd ed. Tsinghua University, Beijing: Springer, 2007.

[191] S. Zhao, "Pulsed ultra-wideband: Transmission, detection, and performance," PhD dissertation, Oregon State University, 2007.

[192] Y. Zheng, M. A. Arasu, K. W. Wong, Y. J. The, A. P. H. Suan, D. D. Tran, W. G. Yeoh, and D. L. Kwong, "A 0.18 μm CMOS 802.15.4a UWB transceiver for communication and localization," in *IEEE International Solid-State Circuits Conference, ISSCC*, 2008.

[193] X. Zhuge and A. G. Yarovoy, "A sparse aperture MIMO-SAR-based UWB imaging system for concealed weapon detection," *IEEE Transactions on Geoscience and Remote Sensing*, vol. 49, pp. 509–18, January 2011.

[194] T. Zwick, C. Fischer, and W. Wiesbeck, "A stochastic multipath channel model including path directions for indoor environments," *IEEE Journal on Selected Areas in Communications*, vol. 20, no. 6, pp. 1178–92, 2002.

[195] L. Żwirełło, C. Ascher, G. Trommer, and T. Zwick, "Study on UWB/INS integration techniques," in *8th Workshop on Positioning Navigation and Communication, WPNC*, April 2011.

[196] L. Żwirełło, M. Harter, H. Berchtold, J. Schlichenmaier, and T. Zwick, "Analysis of the measurement results performed with an ultra-wideband indoor locating system," in *7th German Microwave Conference, GeMiC*, March 2012.

[197] L. Żwirełło, C. Heine, X. Li, T. Schipper, and T. Zwick, "SNR performance verification of different UWB receiver architectures," in *European Microwave Conference, EuMC*, October 2012.

[198] L. Żwirełło, C. Heine, X. Li, and T. Zwick, "An UWB correlation receiver for performance assessment of synchronization algorithms," in *IEEE International Conference on Ultra-Wideband, ICUWB*, September 2011.

[199] L. Żwirełło, M. Hesz, L. Sit, and T. Zwick, "Algorithms for synchronization of coherent UWB receivers and their application," in *IEEE International Conference on Ultra-Wideband, ICUWB*, September 2012.

[200] L. Żwirełło, M. Janson, C. Ascher, U. Schwesinger, G. Trommer, and T. Zwick, "Localization in industrial halls via ultra-wideband signals," in *7th Workshop on Positioning Navigation and Communication, WPNC*, March 2010.

[201] L. Żwirełło, M. Janson, C. Ascher, U. Schwesinger, G. F. Trommer, and T. Zwick, "Accuracy considerations of UWB localization systems dedicated to large-scale applications," in *International Conference on Indoor Positioning and Indoor Navigation, IPIN*, September 2010.

[202] L. Żwirełło, M. Janson, and T. Zwick, "Ultra-wideband based positioning system for applications in industrial environments," in *European Wireless Technology Conference, EuWIT*, September 2010.

[203] L. Żwirełło, L. Reichardt, X. Li, and T. Zwick, "Impact of the antenna impulse response on accuracy of impulse-based localization systems," in *6th European Conference on Antennas and Propagation*, March 2012.

[204] L. Żwirełło, T. Schipper, M. Harter, and T. Zwick, "UWB localization system for indoor applications: Concept, realization and analysis," *Journal of Electrical and Computer Engineering*, 2012.

[205] L. Żwirełło, J. Timmermann, G. Adamiuk, and T. Zwick, "Using periodic template signals for rapid synchronization of UWB correlation receivers," in *COST 2100 TD(09)848*, May 2009.

Index